P9-BYJ-073

Dangerous Diagnostics

DANGEROUS DIAGNOSTICS

The Social Power of Biological Information

DOROTHY NELKIN

LAURENCE TANCREDI

BasicBooks
A Division of HarperCollins*Publishers*

Library of Congress Cataloging-in-Publication Data

Nelkin, Dorothy.
Dangerous diagnostics • the social power of biological information / Dorothy Nelkin, Laurence Tancredi.
p. cm.

Includes index.
ISBN 0-465-01573-5 (cloth)
ISBN 0-465-01572-7 (paper)
1. Medical screening—Social aspects. 2. Human chromosome abnormalities—Diagnosis—Social aspects. 3. Neuropsychological tests—Social aspects. I. Tancredi, Laurence R. II. Title.
RA427.5.N45 1989
306.4'61—dc20 89-42509
CIP

Printed in the United States of America
Designed by Vincent Torre
91 92 93 94 C/CW 9 8 7 6 5 4 3 2 1

CONTENTS

PREFACE

THIS BOOK is intended to generate discussion and debate about the potential uses and abuses of the biological tests that are emerging from research in genetics and the neurosciences. Frequently, technologies that are closely tied to important scientific advances are developed and implemented without adequate public debate. Deeply rooted beliefs about the moral and political neutrality of science tend to preclude debate about the interests served by scientific developments. Hence, the questions of power and social control intrinsically associated with contemporary scientific advances are only weakly articulated.

Our study is a form of technology assessment, but one that focuses less on the technical and commercial possibilities of particular scientific advances than on their inherent social and political implications. Assessments necessarily have a futuristic quality, requiring projections about how technologies are likely to be employed. New technologies are usually applied in ways that reinforce old convictions and support existing institutional practices. Reasonable projections, therefore, must be based on some understanding of the institutional context in which technologies will be applied.

To develop such understanding, we have reviewed the relevant scientific and professional literature from several disciplines (law, psychiatry, the neurosciences, genetics, and the social sciences) and gathered administrative documents (trade journals, memoranda, and so forth) from various institutional sectors. We have surveyed the press and popular magazines to explore the public's understanding, and expectations, of biological testing. And we have gathered information about incidents, cases, and events that we feel cumulatively show up patterns in the use of biological tests.

Gathering material for a book that crosses diverse disciplines and institutional arenas requires extensive cooperation from many people. We would like to thank those who helped us cull information and check the accuracy of our work. These include Paul Billings, Roger Bulger, Gerald Coles, Elaine Draper, Rochelle Dreyfus, Matthew Edlund, Stephanie Goss, Barbara Heyns, Stephen Hilgartner, Robert Proctor, Will Provine, Lola Romanucci-Ross, Gloria Ruby, Margery Shaw, Deborah A. Stone, Nora Volkow, Diane Chapman Walsh, David Weisstub, and Chris Zarafonetis. We also obtained many useful research leads from the minutes of the meetings of the Genetics Study Group in Cambridge, Massachusetts.

We owe very special thanks to Cecile Ervin, Susan Lindee, and Judith Watkins for their extensive research assistance. We also want to thank Betty Goodrum and Paula Obsta for helping to prepare the manuscript.

Our research was facilitated by a grant from Medicine in the Public Interest. Dorothy Nelkin's appointment as the Clare Boothe Luce Visiting Professor at New York University, supported by the Henry Luce Foundation, provided the time to write. NYU's Department of Sociology and Dean C. Duncan Rice provided resources and collegiality. Laurence Tancredi's faculty positions at the University of Texas Medical School and in the Health Law Program at Houston provided needed time, resources, and access to research materials.

Dangerous Diagnostics

Drawing by D. Fradon; © 1987 The New Yorker Magazine, Inc.

CHAPTER 1

The New Diagnostics

IN FRONT of a drive-through testing center on a busy highway, a sign advertises the tests that are given there: "emissions, drugs, intelligence, cholesterol, polygraph, blood pressure, soil and water, steering and brakes, stress, and loyalty."[1] In this single image, a *New Yorker* cartoonist has identified and described a ubiquitous trend in American society: the increasing preoccupation with testing. Like fast food or gasoline, testing is offered as a commodity; one can drive up, be tested, and drive away. The center does not differentiate people from machines; both are objects that can be reduced to parts, examined, and assessed. Nor does it distinguish people's physical health from their behavior. Blood pressure, deviance, intelligence, integrity, political loyalty—all can be routinely tested, just like the brakes of a car. Finally, most of the tests available in this drive-in station are intended not simply to diagnose manifest symptoms of illness or malfunction but rather to discover the truth behind appearances, to detect conditions that are latent, asymptomatic, or predictive of possible future problems.

These are precisely the characteristics of the "new diagnostics"—tests that are emerging from research in genetics and the neurosciences. This research is uncovering the inner structure and function of the body, as well as the inherited qualities that seem to influence the course of life from childhood to old age. Discoveries in these fields are accompanied by diagnostic technologies that have expanded our ability to detect subtle biological differences among individuals and to predict many diseases before symptoms appear.

Technologies such as Computer Electroencephalograms (CEEG), Magnetic Resonance Imaging (MRI), and Positron Emission Tomography (PET) can create detailed images of the brain to detect small structural or functional abnormalities that may indicate future neurological or behavioral conditions. Such devices will one day be able to diagnose behavioral disorders before their symptoms are expressed, as is currently the case with Huntington's disease, where PET shows metabolic abnormalities predictive of later disease development.

Scientific advances in genetics are yielding methods of scanning the sequences of DNA that form the basis of our biological inheritance. With these methods, the "markers" for defective genes can be detected. To date, hundreds of genetic markers indicating predisposition to a growing number of hereditary diseases have been identified. As more genes are identified, tests will be able to indicate predisposition to mental illness, hyperactivity, alcoholism, or addiction, and even a range of personality traits.

Many of the more sophisticated genetic and neurological tests are still limited to experimental investigation. Yet the new diagnostics are providing theoretical models that explain complex human behavior in simple biological terms. As new tests become available to predict common disorders with a genetic component, they are expected to become routinely incorporated into medical and psychiatric practice.

Some physicians fear that reliance on statistically based evaluations will lead to the neglect of important "soft" data and divert attention away from the importance of environmental influences on human behavior and disease.* Despite these reservations, more and more articles appear in medical journals explaining diseases in genetic terms, and the results of the new testing are increasingly included in patients' medical records.[3] Genetics, in the words of the president of the Society for Human Genetics, has "made inroads into the medical mind."[4] Commercial firms have encouraged these developments. Anticipating a massive market in genetic tests, biotechnology companies are making heavy investments in techniques to identify genetic markers and to create diagnostic kits. "The future lies in genetic tendencies," says the scientific director of a major biotechnology firm.[5]

Similarly, the use of improved imaging technologies, such as CEEGs, developed from the neurosciences is becoming common in psychiatric practice. In his 1987 presidential address to the American Psychiatric Association, Dr. Robert Pasnau observed the "remedicalization of psychiatry . . . as scientific advances in the neurosciences affect psychiatric diagnoses and treatment."[6] Conversely, influenced by the research on the neurophysiological bases of behavior, neurologists have turned increasingly to the study of behavioral problems, even creating a subspecialty of "behavioral neurology" grounded in biomedical tests. These scientists expect new imaging devices to have many applications for behavioral disorders, lending insight into the fundamental processes of brain function: "Pre-symptomatic detection of psychiatric disease will be routine," claims the National Institute of Mental Health. "Subjects at risk for alcoholism, schizophrenia, or depression may be identifiable well before the onset of clinical symptoms."[7] Reacting to such predictions, psychiatric hospitals are purchasing imaging technologies ranging

*Alvan Feinstein points out in his provocative book *Clinimetrics*[2] that some of the most cogent distinctions in diagnosis, prognosis, and therapy depend on characteristics such as the severity of symptoms, the level of actual functional capacity, and changes in the condition of a patient over time that are not amenable to statistical analysis. He and others prefer to rely on the informed intuition of the individual than on the diagnosis of a machine.

from moderately inexpensive CEECs to PET equipment costing over $4.5 million.[8]

The medical system is the core around which other institutions build their use of new diagnostic techniques. Their acceptance in medical settings is encouraging their use outside the clinical context: in insurance companies, in schools, in the workplace, in the courts. The accumulation of diagnostic information about individuals can indicate preventive actions or therapeutic procedures. However, nonclinical institutions may use these tests in ways that the medical profession did not intend, and sometimes with devastating results.

Imagine, for example, a small electronics firm planning to manufacture an intricate navigational system. Part of the system must be assembled manually, and the company must carefully select an individual for the arduous training this new job requires. One of the company's employees seems ideal. Dependable, bright, and particularly quick with her hands, she is thirty-five years old and has worked for the company for eight years. This company routinely uses the latest diagnostic tests to screen its employees, and has her biological profile on record. The company physician does a computer scan of it, searching for disabilities that could affect her motor coordination in the future. He discovers an unexpected problem: the woman's DNA markers show, with a high degree of certainty, that she will develop Huntington's disease, an inherited degenerative neurological disorder that results in loss of motor control, depression, personality changes, and death. Symptoms may begin to develop when she is about forty, and soon after she may be unable to work. Her condition would eventually become debilitating and require costly medical care. Thus she would be an economic drain on worker's compensation and the corporation's health insurance and long-term disability policies. From the employer's perspective it would hardly be rational to promote and to train this person. But for the individual, the consequences of this information are disastrous.

Her employer might begin to question her ability to function in her current job. If she is terminated, who else would employ her? And who would pay her costly medical bills?

Sophisticated diagnostic tests serve many useful and humane purposes: in clinical settings they may point the way to particular therapeutic measures. They can provide families the opportunity to avoid the anxiety and cost of bearing a child with an untreatable disease; they can identify potential health or behavioral problems for remedial or preventive action. In nonclinical contexts, they can help in the early recognition of learning-disabled children, protect vulnerable workers from exposure to harmful toxic substances, and provide solid evidence for legal decisions about a person's criminal responsibility or competence to stand trial. The language used to describe diagnostic techniques speaks mostly of such benefits. The tests emerging from research in the neurosciences will "generate clinical successes" and provide "answers to disabling mental illness," claims the National Institute of Mental Health.[9] "New genetic clues to heart disease, cancer, AIDS and other killers could save your life," reports a journalist.[10] "We'll achieve the ideal in medical care, the prevention of disease," predicts the director of a biotechnology firm.[11]

Yet information from tests is not always beneficial or even benign, for in many cases nothing can be done to prevent the predicted disease. What will happen to a twenty-year-old who discovers that he is likely to develop a fatal disease in middle age? The genetic flaws detected by tests will not necessarily translate into functional impairment; yet knowing about the potential problem without being able to prevent it will be a source of extreme anxiety for him. Moreover, it could subject him to considerable discrimination. He may be denied a job, and his insurance costs will surely increase. Even if something can be done to prevent the manifestation of a predicted condition, awareness of the predisposition can be used in ways that harm the individual. A diagnosed genetic vulnerability to heart disease may encourage a preventive lifestyle, but the prediction itself could affect a person's career. After all, tests have often been abused, serving, for example, as a means to justify racial or gender bias, to legitimate arbitrary exclusionary practices, and to enhance institutional power with little regard for the rights or personal fate of individuals.

The prerogative to test has long been recognized as a source of power and social control. Michel Foucault saw the examination as a strategy of political domination, a means of "normalization." In *Discipline and Punish* he described the examination as "a normalizing gaze" that "introduces the constraints of conformity . . . [that] compares, differentiates, hierarchizes, homogenizes, excludes." On the extension of testing throughout society, Foucault observed: "The judges of normality are present everywhere. We are a society of the teacher-judge, the doctor-judge, the educator-judge, the social worker-judge. . . . We are entering the age of the infinite examination and of compulsory objectification."[12]

Foucault wrote of pedagogical tests. Others, such as Walter Reich, a psychiatrist at the National Institute of Mental Health, have developed similar analyses of psychiatric tests. Tests, they say, are often used to reinforce political hierarchies or social values. Reich warns of the "nonmedical beauty of diagnosis . . . [its] powerful, varied, and unrecognized role in the lives of all persons," and its "susceptibility to misuse."[13] He cites the abuse of psychiatric testing in the Soviet Union as an extreme example, but observes that psychiatric tests are also routinely used to explain behavior that violates social norms, and to label and therefore discredit those who threaten the power of existing institutions.

Our analysis focuses on biological tests. Our point here is not to debunk diagnostic techniques, but rather to generate discussion about the development and use of the new diagnostics by exploring the social, cultural, and institutional contexts in which they are applied.

The Cultural Appeal of Testing

The increased preoccupation with testing reflects two cultural tendencies in American society: the actuarial mind-set, reflected in the prevailing approach to problems of potential risk, and the related tendency to reduce these problems to biological or medical terms.[14] Actuarial thinking is designed to limit liability. It requires calculating the cost of future contingencies, taking into account expected losses, and selecting good risks while excluding bad ones. The individual must therefore be understood actuarially, that is, with reference to a statistical aggregate. In this context the information derived from tests becomes a valuable economic and political resource.

The actuarial mind-set thrives on information about the health, habits, and behavior of individuals. Personal data are the grounds for organizing access to credit, monitoring taxation, managing school registration, implementing criminal investigations, and planning for future health care needs. The gathering of personal information—credit rating, health characteristics, and behavior—by government agencies, employers, and schools has increased dramatically in the past twenty years.[15]* Testing is

*In 1970 a U.S. Senate subcommittee held public hearings to assess concerns about a "dossier dictatorship." Testimony revealed that the private life of the average American was already the subject of 10 to 20 dossiers in government and private agency computer files. "Americans are scrutinized, measured, watched, counted, interrogated by more government agencies, law enforcers, social scientists and poll takers than at any time in our history."[16] While most data are gathered for socially desirable ends, the public hearing highlighted their potential danger in a society where facts have political and economic value. In the 1980s the ability to gather and store information has greatly increased. Sophisticated techniques of computer profiling and monitoring and more refined testing and screening methods facilitate data collection. Laser techniques allow information to be stored efficiently on small identification cards which could contain an individual's genetic profile in easily accessible form.

part of this trend. Screening prospective employees for drug use has become much more common despite concerns about the accuracy and legitimacy of such tests. There is relentless pressure to test for AIDS despite the discriminatory implications. And the use of standardized tests in schools is growing despite questions about their validity as a measure of intelligence and predictor of performance. Indeed, faith in "facts," in the numbers derived from testing, has obscured the uncertainties intrinsic to such diagnostic tests, and they are widely accepted as neutral, necessary, and benign.

Just as actuarial reasoning supports reliance on testing, so too does the tendency to reduce complex behavior to measurable biological dimensions—that is, dimensions that can be revealed through a test. Biological reductionism is part of the pervasive tendency to medicalize social problems.[17] The medical profession has long interacted with various social institutions to assist in and legitimize social policy. Physicians and psychiatrists interact with the law by reporting venereal disease or gunshot wounds; with schools by evaluating absenteeism or learning problems; with industry by judging responsibility for illness or the capacity to work; with hospitals by establishing patient competence in cases of controversial administrative decisions; with the military by authorizing deferments; and with the courts by determining the disposition of prisoners or their competence to stand trial. As medical testing has become more refined and accurate, these institutions have placed more reliance on medical judgment to define the boundaries of "normal" behavior and thereby to identify competence, deviance, or capacity to work.

In its contemporary manifestation, this medicalization has incorporated notions of biological fitness or perfectibility.[18] It is assumed that there is an ideal of biological normality or perfection against which individuals can be measured, that complex human behavior can be reduced to biological or genetic explanations, and that behavioral problems can be attributed to biological determinants with minimal reference to social or environmental influences.

A great deal has been written about the biological determinism of the late nineteenth and early twentieth centuries and the prevailing assumption that economic and social differences between groups (races, classes, and sexes) were based on inherited distinctions.[19] IQ tests, psychological testing, and even the measurement of facial and cranial characteristics were the means to justify these beliefs.[20] Until World War II, eugenicists dominated the field of human genetics. *Eugenics*, a word coined by the English scientist Francis Dalton in 1869, refers to the "science" of improving the hereditary qualities of the human race. Eugenicists sought policies to encourage breeding among those deemed socially worthy ("positive eugenics") and to discourage breeding among the socially disadvantaged ("negative eugenics"). An active American eugenics movement, for example, encouraged marriages among "fitter families" and promoted large-scale projects to sterilize persons defined as "unfit." Testing gave their policies scientific legitimacy. Using biology to justify class and racial biases, eugenicists contributed to legislative decisions on the exclusion of immigrants, access to jobs, and educational opportunities. They explained in terms of genes virtually all the characteristics of immigrant groups, misfits, criminals, and the mentally deficient.

The widespread acceptance of biological reductionism declined after the Nazis implemented eugenic ideas in the atrocities of World War II.[21] After the war, it was no longer politically acceptable to look to biology as a guide to social policy.[22] Genetic explanations of human behavior were replaced by cultural or psychological analyses, only to re-emerge in the scientific discourse of the 1960s, as scientists once again expounded the social meanings inherent in their work. Writing a history of his discipline in 1987, German geneticist Benno Muller-Hill observed that "the rise of genetics [was] characterized by a gigantic process of repression of its history."[23] Scientists are oriented to the present. Less than twenty years after the Nazi atrocities, they have, in effect, obliterated history. Thus, in 1963 chemist Linus Pauling stated that parents who know they are carriers of the cystic fibrosis allele and who continue to produce defective children "add to the amount of human suffering and should feel guilty for their actions."[24] In 1970, the Committee on Science and Public Policy of the National Academy of Sciences published a survey of the life sciences with the conclusion that eugenics is a means to "expand the human potential" to produce a "healthier society."[25] And in the same year, Bentley Glass, in his retirement address as president of the American Association for the Advancement of Science, called for "the use of the new biology to assure the quality of all new babies. . . . No parent will have the right to burden society with a malformed or mentally incompetent child."[26] He predicted that recessive hereditary defects would be detectable in carriers, who should be warned against or prohibited from bearing offspring.

Critics responded, warning of the potential abuses of a science that presumed to predict biological destiny.[27] A growing feminist literature also challenged the powerful biological assumptions underlying many aspects of social policy. But in the 1980s, the quickening pace of discoveries in the genetic sciences has inspired frequent references to the genetic basis of mind, learning, and behavior. In the long debate over the relative influences of "nature and nurture," the balance seems to have shifted to the biological extreme. Some scientists have claimed that the nature-nurture controversy is dead, declaring nature the winner.[28] And, accordingly, they have become less and less reluctant to discuss its implications for social policy.

New theories of biological determinism are based on far more sophisticated scientific knowledge than those of the nineteenth century. This knowledge has generated eugenic-like themes in scientific discourse. Some scientists continue to extrapolate from individual differences to the characteristics of groups, and still write about the relationship of race to IQ, the inherent differences in the mathematical abilities of men and women, and the special susceptibility of blacks to harm from chemicals in the workplace.[29] There is even a small organization, founded in 1982, called the Eugenics Special Interest Group that tries to "provide a communications network for all people committed to enhancing human genetic quality; to develop sound, innovative projects to produce eugenic benefit; . . . and to enlighten the public about the unique potential of eugenics with the intention of ultimately influencing public policy."* For the most part, however, the "new eugenics" has avoided generalizations about class and race, focusing instead on the individual benefits that follow from genetic research.

Although the old eugenic generalizations have been cast off, the logic behind them persists, refueled by evidence from diagnostic tests and justified in terms of efficiency, effectiveness, and cost. Thus some geneticists suggest the social importance of improving the "gene pool." For example, geneticist Margery Shaw, convinced that every Mendelian genetic trait will eventually be diagnosed prenatally, has asserted that: "The law must control the spread of genes causing severe deleterious effects, just as disabling pathogenic bacteria and viruses are controlled."[30] She argues that parents may be liable for failing to respond to information about potential genetic disorders by controlling their reproduction, and that the police powers of the state could be employed to prevent genetic risks. Other geneticists assume that families informed of genetic problems will voluntarily eliminate defective fetuses.[31] References to the "pollution of the gene pool," "genetically healthy societies," and "optimal genetic strategies" are beginning to appear in the scientific discourse.[32] The language of geneticists reveals their expectations. They have

*ESIG is part of MENSA. Based in Tennessee, it publishes an occasional magazine called the *Eugenics Bulletin*.

called the large-scale project to map the human genome* a "quest for the Holy Grail" and an effort to create the "Book of Man." The computer program that generates the genome is called GENESIS.

The popular culture has supported and publicized scientific generalizations that reduce complex problems to simple biological explanations. In the late 1970s, following E. O. Wilson's *Sociobiology*,[33] the press used sociobiological concepts to explain the most diverse aspects of human behavior. Selfishness, according to *Psychology Today*, is "built into our genes to ensure their individual reproduction." In 1980 *Time*, writing of the "gender factor in math," suggested that "males may be naturally abler than females." *Discover* reported that male superiority in mathematics is so pronounced that "to some extent it must be inborn." Covering the Baby M case in 1987, a story in *U.S. News and World Report* called "How Genes Shape Personality" proclaimed that "solid evidence demonstrates that our very character is molded by heredity." The article, therefore, questioned whether Baby M's future really hinged on which family would bring her up.[34] By the mid-1980s, terms such as "genetic vulnerability," "genetic predisposition," and "genetic tendency" were part of the common parlance of both academics and the press. The 1985 publication of *Crime and Human Nature*, James Q. Wilson and Richard J. Herrnstein's review of the theories on the biological basis of criminal behavior, became a media event and the subject of lead articles in popular magazines and newspapers.[35]

The possibility of genetic prediction is clearly attractive to the many people who are preoccupied with tracing their heritage in order to discover their predisposition to disease. Dubbed the "worried sick" by Harvard psychiatrist Arthur Barsky, such people have concerns that are one aspect of the growing tendency of

*The *genome* is the full range and variety of genetic material present in the chromosomes of a particular organism.

healthy people to focus on potential medical risks.[36] Indeed, one popular magazine advises its readers to find out their medical histories by developing a "genogram," that is, a family tree of the physical and mental ailments of their relatives.[37]

Metaphors of the Body and Mind

The popular discourse about behavior and disease is being shaped by current research trends in genetics and the neurosciences. Centered on the importance of biology in predicting future behavior and health, this research appears to reduce human behavior to objective biological determinants that are waiting to be discovered. It presents an image of the body and mind as machinelike "systems" that can be visualized on a computer screen and understood simply by deciphering a code. This metaphor underlying the development of the new diagnostic tests has shaped how tests are interpreted and their results employed.

Metaphorical constructs have commonly been used to conceptualize the body and mind. The particular metaphors that have shaped our understanding of the body have changed over time, following scientific and medical developments, but even as they evolved, they have continually reinforced the view of the body as a mechanical system. The advances in organ transplantation during the 1960s, for example, projected an image of the body as a set of replaceable parts. This was a popular metaphor; the idea of replacing arms, legs, liver, spleen, reproductive organs, and even the brain captured the public imagination.

Another metaphor began to emerge as early as the 1930s, when antibiotic medications created the image of the body as a chemical system. Developments in molecular biology in the late

1960s and the 1970s that elucidated the chemical basis of cellular activity pushed this metaphor to the forefront. Then the genetics revolution, with the disclosures of DNA and RNA, created the image of the body as a chemical factory, an entity composed of many cells, all occupied with manufacturing "chemical building blocks."

Metaphors describing the relationship of the brain to the mind (or mental activity) developed along similar lines. During the 1940s and 1950s, research on neuronal activity and electrical impulses shaped a perception of the brain as a very complex system of electric wires. Outside stimuli entered the senses and were conducted through intersecting nerves to various sites in the cerebral cortex. Then the growing role of computers in the area of medical diagnosis and treatment shaped a metaphor of the mind as a computerized system working on logic similar to the binary system basic to computer technology. The "hardware" equivalent of the computer was the biological tissue of the brain; the "software" was the brain's mental activity, that is, the program that generates ideas and associates images, sensations, and thoughts. Research on artificial intelligence soon revived interest in neural network theory, combining features of the computer and wired-network metaphors to explain how humans learn, perceive, and remember.[38]

For bioscientists, metaphors of the mind changed considerably in the mid-1970s. Discoveries such as the role of the hypothalamus in producing activating hormones strongly indicated the influence of the glandular-producing capacity of the brain on shifts in mood and emotions. This suggested that the brain was more than just a computer; it was a biological entity similar to other glands operating in the endocrine system of the body. This metaphor, influenced by images from the diagnostic technologies that visualize chemical and biological processes, conceptualizes the brain as a producer of neurochemicals, from those that affect the transmission of information to those that are implicated in serious psychiatric illnesses such as schizophrenia and the affective disorders.

These metaphors of the body and mind have, in effect, objectified the person, who becomes less an individual than a set of mechanical parts or chemical processes that can be calibrated and well defined. This objective image of the person has encouraged the use of biological tests as means of classification and as instruments of control.

Institutional Uses of Tests

Tests are in the first instance a means of defining appropriate therapeutic measures.[39] But they are also a way to create social categories (of "learning-disabled" children, "high-risk" employees) in order to preserve existing social arrangements and to enhance the control of certain groups over others. As Foucault observed, tests are instruments of control. The power to define the normal can impose standards of conformity, while the ability to measure individual deviations can justify classification and hierarchy. Tests have long been used to measure competence, to define deviance, and to exclude those who appear less desirable to a group's social or economic goals. Technologies that tap into biological understanding of how the body functions and how it can be expected to function during the course of an individual's life are an extension of earlier pedagogical and psychiatric tests. And, like these tests, they serve as gatekeepers, controlling access to employment, hospitals, and schools, and allowing organizations to shape their clients as a projection of their own economic and administrative needs.

For many organizations the use of predictive tests is economically rational. But organizations operate in a context of constant tension between their economic needs and the rights of their clients. While biological tests enhance institutional control, they can also conflict with social considerations of civil liberties, human integrity, or personal privacy. They can bear on people's economic interests, the cost of their insurance, their access to jobs and educational opportunities. They can affect personal well-being, how people are labeled, their self-conceptions, their exercise of free will. They can influence social relationships, leading to stigmatization and discrimination.

In examining the use of tests in specific organizational settings, we will suggest how diagnostic advances and their underlying scientific assumptions affect the boundaries of what is viewed as acceptable behavior; how they delineate and define classes of what is or is not "deviant" or diseased. We will examine the process through which an increasing number of people can be marginalized, excluded, or stigmatized as a result of predictions based on biological tests. We will look at the expanding role of the science-based professions in providing more reliable diagnoses of physical and psychological disorders, giving greater power to the company physician, the school psychologist, the genetic counselor, and the neuropsychiatrist in the courts. And we will explore how the new diagnostics are becoming a means for organizations to extend their control beyond the institutional arena (of work, of education)—to the personal life of their clients and to their future possibilities.

We are especially concerned about the use of tests in situations where the interests of an organization in maintaining order place a high value on conformity, for example, in the case of overactive children in schools. We will explore the use of tests to predict disease in asymptomatic persons when the meaning of test results is uncertain, for example, in some pre-employment examinations. We will focus on situations where prediction of future illness on the basis of tests creates the capacity for exclusion—from insurance or from work—or creates impossible choices between work and health. And we will explore situations where the refined capacity to locate subtle abnormalities can reinforce subjective labels, affecting, for example, decisions about competence to stand trial.

There are, of course, clear-cut diagnostic categories—true illness is not to be denied—but illness and deviance can also be social constructs. The boundaries of the "normal" or the "healthy" are often fuzzy. At what point, for example, should a person with borderline behavior be labeled learning-disabled or mentally ill? Boundary problems are most obvious in defining mental illness. "Suspiciousness" could be defined as a paranoid reaction or an understandable response to environmental pressures. Social constructs can be important even in defining physical disease. In parts of Africa the sickle-cell trait is a normal adaptive reaction to malaria; in the United States it is viewed as evidence of disease. When interpretive margins are fluid, they can easily be manipulated by diagnostic evaluations cloaked in the neutral garb of science. For normative assumptions about scientific objectivity enhance the power of diagnosis and conceal the values embedded in many tests.*

Our purpose, then, is to demystify the new diagnostic technologies by revealing the organizational pressures and preconceived values shaping their applications and by opening debate on their social implications.

*A growing body of empirical studies in the social construction of science has challenged conventional views about the value neutrality of science. The social constructivists document how scientific facts emerge from the interaction between laboratory scientists and those state and corporate interests that translate science into technology, and they stress the importance of political pressure, economic interests, and social objectives in shaping both the construction of "facts" and the public acceptance of claims to scientific knowledge. They also suggest how science becomes a means of persuasion, a way to legitimate social policy and institutional practices as based on "fact," and therefore nonrefutable and independent of social context.[40]

CHAPTER 2

Defining Diagnosis

FOR MOST of medical history, diagnosis has been limited to measuring the external manifestations of disease. Only by inference could these manifestations be linked to the deeper structural problems underlying the presence of disease. The general physical examination, for years the standard forum for the discovery of unsuspected illness, relied on the senses of the physician to perceive such signs of disease as erratic heartbeats or lung movement. Few tools were available to confirm the physician's intuitive impressions about a patient's condition.

The use of diagnostic technologies in medical care began in the late nineteenth century with the introduction of the stethoscope and the thermometer into regular medical practice. These simple technologies radically changed the way physicians defined disease. Prior to this period, disease was defined less by the type of symptoms exhibited than by the "natural" characteristics of each patient. Disease was believed to affect each patient differently, depending on age, gender, ethnicity, socioeconomic position, and moral status.* Once able to quantify patients' symp-

*The idea of a "natural state of the individual" was rooted in the principle of specificity, defined as an individualized match between medical treatment and the specific characteristics of each patient.[1]

toms, physicians began to focus on standardized symptomatology, defining disease according to fixed norms. This change had fundamental implications for both medical theory and therapy, for it shifted attention from the symptoms reported by patients to those that were accessible through the intervention of technology. As the perceived value of quantitative evidence steadily increased, so too did the appeal of machines that could suggest the cause of human suffering.

In the early part of this century, physicians could take blood pressure readings with a sphygmomanometer, measure the pulse, or use the most rudimentary devices such as a tongue depressor to examine a patient's throat.[2] The medical use of X-rays, which began shortly after their discovery in 1895, demonstrated that diagnostic instruments could visualize the body's internal structure. The usefulness of the X-ray encouraged the development of quack as well as legitimate electronic and mechanical devices. For example, the Drown Instruments promised to recognize disease long-distance, through radio waves; the Radioclast was to measure "vibration rates" in the body; the Orgone Energy Accumulator was promoted as a means to identify and cure sexual dysfunction; and the Sonus Film-O-Sonic was actually supposed to diagnose and treat cancer by playing the song "Smoke Gets in Your Eyes."

Still in use during the 1960s, these instruments were attacked by the American Medical Association (AMA) in a 1968 pamphlet as "fake health machines in the hands of quacks." The AMA insisted that there existed no machine "capable of diagnosing or treating different kinds of diseases simply by turning dials and applying electrical contacts to the body." The machines were "medically worthless," said the AMA pamphlet, though "impressive in appearance in order to fool the gullible."[3] But the medical establishment did, in the 1940s, approve the use of the fluoroscope, a device using X-rays to project shadows of bones or other internal body structures onto a fluorescent screen. Despite its limited value as a diagnostic tool, the fluoroscope became a symbol of scientific sophistication, used by many private physicians until they recognized it as a harmful source of radiation.

The popularity of fluoroscopes reflected the appeal of machines that could "image" what is actually occurring within the body. By the 1970s research in radiology and neurobiology enhanced by advancements in the computer was providing the basis for new diagnostic techniques that allowed noninvasive visualization of both the internal structure and the function of the body and the brain. At the same time scientific understanding in the field of genetics also led to new developments in diagnostic testing. In both cases, the new technologies represented a conceptual change in diagnostic capabilities.

Like all diagnostic techniques such as IQ tests or tests of body temperature, both genetic and imaging technologies function by establishing a statistical definition of "normal," which becomes a standard against which to measure individual differences. Deviations from statistically defined standards are defined as abnormal or deviant. In addition, the new diagnostics share several related characteristics.

First, they reveal the biological and subcellular substrates from which a person's physical or psychological characteristics emerge. Able to detect very early biological changes, they enhance diagnostic predictability, allowing anticipation of problems that might not be visibly expressed in overt symptoms for years.

Second, they can detect ever more minute individual differences with increased precision. Just as sophisticated detection technologies have enhanced our ability to measure carcinogens in food products with greater sensitivity, so improved diagnostic techniques have refined our ability to identify subtle deviations from the norm. In both cases advances in testing have redefined what is normal and expanded the number of products—or persons—identified as problematic.

Third, these diagnostic technologies are useful not only for clinical diagnosis—that is, for providing an explanation of an individual's health status—but also for screening, where the purpose is to identify from a large population those individuals who in some way deviate from the statistically derived norm. Tests that detect those potentially at risk are, of course, intended to be useful in a range of social and economic institutions outside the clinical medical context. This has encouraged powerful commercial interests to develop and extend the application of such tests.

Finally, even when limited to experimental use, the new biological tests have an aura of precision and scientific objectivity that enhances their credibility. Thus they are framing the professional and popular discourse about social and individual problems, shaping both institutional practices and social policies.

The Uses of Diagnosis

Historically, the development of diagnostic techniques has outpaced the ability of medical practitioners to provide appropriate treatment. Until the twentieth century, clinicians had focused most of their attention on refining the art of diagnosis, for there were few treatments available for the many existing diseases.[4] Lacking therapeutic options, they emphasized the importance of describing their patients as accurately and scientifically as possible. Necessarily, the process of discovery became an objective in itself. The value of the test rested less on its therapeutic implications than on its contribution to the understanding of a problem and the refinement of medical evaluation.

This attitude has continued to the present. While it has become possible to acquire more and more information about bodily processes, this does not necessarily lead to benefits beyond ordinary care and reassurance. Therapeutic options are often limited. Studies of the influence of diagnoses on therapy suggest that the information produced by sophisticated diagnostic technologies affects the course of therapy in only 8 to 30 percent of cases.[5] Even with dramatic therapeutic advances, many diseases that can be identified cannot be cured.[6] Thus, while the information gathered through clinical diagnosis is intended to guide therapeutic strategies, understanding the biophysical system for purposes of prognosis and prediction remains a primary goal.

In contrast, the use of diagnostic tests for screening has always been directly linked to social or medical intervention, through therapy, prevention, or exclusion. The testing of asymptomatic people began in the late nineteenth century, when progressive reformers promoted the idea that adults should have regular precautionary medical examinations. Later the growing use of the automobile, which needed regular checkups, suggested that the human machine, too, could benefit from regular inspection.

The military employed diagnostic screening on a large scale to sort out the "probably well from the probably sick" during World War I; that is, to discover those who, on medical grounds, should be disqualified from the draft.[7] Of the 2.7 million men called into service, 47 percent were found to have physical impairments that went previously undetected. Since many of these impairments could have been prevented, the promotion of preventive health examinations, often for syphilis, became a major objective of public health organizations in the 1920s.[8] The initiative came primarily from consumers, but physicians too took an interest in mass screening programs, and the AMA endorsed regular examinations for those "supposedly in health."

Interest in preventive health declined after the 1920s when the control of disease through mass screening appeared economically impractical. It revived again in the late 1950s and 1960s when the government supported multiphasic screening programs as part of its general public health policy. These programs, consisting of routine urine and blood tests and, in some circumstances, X-rays and electrocardiograms (EKGs), became possible on a large scale with the introduction of efficient automated laboratory analyzers.

The usefulness of particular technologies for screening large populations rests on practical considerations: the technology must be easily applied and the data efficiently recorded; each individual test must be relatively inexpensive or the total cost would be astronomical. For example, drug-abuse screening procedures (through enzyme-mediated immunoassay tests or radio immunoassay tests of urine) are easy to administer and inexpensive, costing between $7 and $20 per specimen. They are therefore a cost-effective way to screen large numbers of prospective employees.

The growing availability of computers and automated diagnostic systems has greatly encouraged the expansion of screening in many nonclinical contexts—for example, in the military to test for drug abuse, in prenatal clinics to test for genetic disease, and in the workplace to define the health status of prospective employees.

Understanding the differences between diagnostic screening and clinical diagnosis is important in considering the use of the new technologies. Within the clinical context, physicians use diagnostic technologies to obtain data about an individual's physical or mental status. When a patient has specific complaints, the physician tries to find a scientific explanation and to locate a cause. In screening, however, the same technologies become a means of identifying a population of individuals potentially at risk so as to influence their behavior. For example, genetic screening of potential carriers of Tay-Sachs disease (a fatal hereditary disorder of lipid metabolism) is a way to identify those who may perpetuate the trait in order to provide proper consultation prior to pregnancy. Prenatal screening for genetic disorders is a way of providing parents with the option of terminating the pregnancy if the condition is sufficiently serious. And genetic screening in the workplace is a way of identifying those most susceptible to toxic chemicals so as to prevent harmful exposure. In other words, the purpose of screening is to reveal conditions that are not yet manifest in order to intervene by taking preventive measures.

Research Developments and Diagnostic Trends

Genetic tests are currently used for both prenatal screening and the identification of adults who may be carriers of a hereditary disease or who may themselves be at risk. Prenatal tests can detect a wide range of chromosomal abnormalities such as Down's syndrome and genetic abnormalities such as Tay-Sachs disease. Perhaps the most well known form of prenatal screening is amniocentesis. A procedure once rare in obstetrical practice, amniocentesis is now routinely recommended for pregnant women who are in their mid-thirties and older or who have family histories of serious congenital and genetic disease.

Amniocentesis is generally used to detect the presence of abnormalities in the number or morphology of chromosomes in the cell. These would include trisomies such as Down's syndrome, unbalanced translocations, mosaics, or sex chromosome abnormalities. But since it provides a sample of the genome of the fetus, amniocentesis can also be used to detect biochemical abnormalities at the genetic level in the DNA. Genes are the functional unit of heredity. Abnormalities may include a gene duplication, an error in the linkage of the nucleotide, or an absence of a specific nucleotide in the DNA. Up to 180 genetic disorders, including serious conditions such as Huntington's disease and sickle-cell anemia, can be revealed by amniocentesis. And the amniotic fluid can also be subjected to serum hexosaminidase assay to detect the presence of Tay-Sachs disease, and to alpha-fetoprotein measurement to detect neural tube defects.

Chorionic villus sampling (CVS), a new prenatal screening technique, can detect genetic abnormalities in a tissue sample from the embryonic membrane that surrounds a young fetus. Using gene probes or chromosomal analysis, CVS can test a fetus at ten weeks, a stage at which termination of a pregnancy is relatively uncomplicated. It is thus an increasingly preferred prenatal screening technique.[9]

Physicians can also use genetic tests to identify people who are carriers of autosomal recessive genes* for diseases such as Tay-Sachs or sickle-cell anemia. The purpose of such tests is to gather information that will provide reproductive choices for carriers of the gene, allowing them to prevent perpetuation of the disease through their offspring. Newborn infants are also routinely screened for metabolic defects such as phenylketonuria (PKU) or hypothyroidism in order to provide guidance for intervention through either dietary treatments or drugs.[10]

Current research in human genetics allows the early identification of an increasing number of diseases through the detection of a marker, an unusual DNA sequence that is believed to be inherited with a disease-causing gene. To identify markers, scientists use recombinant DNA techniques to cleave the DNA drawn from blood cells into thousands of fragments, called restriction fragment length polymorphisms (RFLPs). These fragments can be radioactively labeled and visualized through photographic techniques to create DNA probes, which then serve as tests to identify the region on a chromosome where a defective gene is located.

*An *autosomal recessive* trait is one that is inherited through both parents. For the disease to be manifest, both of the components or genes that constitute the allele (the specific paired loci on the chromosome) must have the abnormality. If only one gene has the abnormality, the individual becomes a carrier for the disease but will not manifest it.

The DNA must be taken from a family that suffers a high incidence of the genetic disease. To identify the telltale pattern or marker, it is necessary to compare the genes of healthy family members with those who actually have the disease. If all those who have the disease share a particular gene sequence, it can be assumed that this sequence is the marker linked to the disease. In fact, the presence of a marker is highly likely—to an accuracy of about 95 percent—to indicate an abnormal gene. Where the DNA probe identifies flanking markers (those on each side of the suspected gene), the level of certainty may be over 98 percent. And if the DNA probe is directed to a marker that is actually inside the gene responsible for the abnormal condition, there is 100 percent certainty that the individual has that particular genetic sequence.[11]

Using these techniques, scientists have identified at least 350 markers for genetic disease. Many of these markers are for single-gene disorders such as Huntington's disease, Fragile X syndrome, sickle-cell anemia, cystic fibrosis, hemophilia, retinoblastoma, and phenylketonuria. But studies of selected RFLP markers in human families have revealed the linkages for more complex and more common diseases. Evidence suggests that specific genes or groups of genes predispose individuals to some forms of cancer, emphysema, juvenile diabetes, Alzheimer's disease, cleft palate, heart disease, and mental illness. Indeed, the pace of discovery in this field has led to jokes among geneticists about "the gene of the week."[12]

The diagnostic potential of these discoveries has been realized in the case of several single-gene disorders, but these represent only a small portion of all genetically transmitted diseases. The more commonplace disorders are caused by multiple genetic and social factors making it more difficult to show a causative relationship between a specific genetic abnormality and its manifestation. Thus the application of research to the development of diagnostic technologies began with single-gene disorders such as Huntington's.

Victims of Huntington's disease usually develop symptoms between the ages of thirty and fifty, later in some cases. About half of those with one afflicted parent (about 125,000 people in the United States are at risk) will inherit the aberrant gene and are certain to get the disease unless they die prior to its manifestation. When scientists found the genetic marker for Huntington's in 1983, they were reluctant to make the test available to physicians, given the lack of therapeutic options. But in 1987 testing became available for asymptomatic persons from families with a history of the disease.

Adult polycystic kidney disease is another inherited, late-onset, and incurable disorder. Likely to develop symptoms in their forties, people with this disease will eventually suffer irreversible kidney failure. There is no known therapy aside from dialysis or kidney transplantation. Children born to a parent with the gene have a 50 percent chance of inheriting it. A genetic test can provide presymptomatic diagnosis, but there is no way to stop the course of the disease. Those who test positive, however, can choose not to reproduce or to practice selective abortion with the aid of prenatal genetic tests.

Scientists are actively pursuing the origins of more complex, polygenic disorders, that is, those involving multiple genes. For example, to target heart disease they are seeking the genetic defects responsible for atherosclerosis. These include an inherited lack of the protein receptor that admits cholesterol into cells. To test for Alzheimer's disease, scientists are looking for a marker on chromosome 21 that codes for superoxide dismutase. And to test for cancer, scientists are developing ways to detect the presence of particular oncogenes (genes implicated in cancer causation) or the absence of growth-suppressing anti-oncogenes as a way to predict malignancy before the disease is clinically evident.

DNA probes have not yet been developed for most genetic conditions, and, except in a few cases, they are not yet widely used as screening devices. However, work is now under way to map out the entire human genome, providing a precise biochemical map or blueprint of the genetic material of human beings, and identifying the markers associated with nearly all 4,000 of the known genetic diseases. This work is facilitated by a new process called DNA amplification which speeds up the duplication of the DNA strands to allow cell growth, enabling better characterization of new markers. When genetic mapping is complete and the human genome has been sequenced, DNA tests will be an expedient means of predicting the growing number of diseases and behavior problems believed to have a genetic component.

A related set of diagnostic technologies has emerged from research in the neurosciences and from resulting changes in the understanding of how the brain functions. Studies have focused on how the various neurotransmitter systems actually operate at the cellular and regional brain levels to affect specific brain functions. In particular, research has looked into the relationship between the chemicals produced in the brain and mental disorders or psychological states. The next twenty years will bring about expansion in the number of biochemicals that are identified in the brain and greater understanding of the link between various neurotransmitter systems and human behavior. Current knowledge is forming the basis of increasingly sophisticated diagnostic technologies.

Computer Assisted Tomography (CAT) provides structural images of the brain and body based on the different densities of the human tissue. Magnetic Resonance Imaging (MRI) provides structural images based on the magnetic properties of certain elements such as phosphorous in the tissue. Computers translate these signals into an image on a screen or printout, allowing the detection of tumors or deviations from normal structures. While both CAT and MRI identify subtle structural variations, Positron Emission Tomography (PET) traces the metabolism and neurochemical characteristics of the brain under various conditions, and therefore can be used to assess the actual working of the brain.

Currently PET is the main imaging device for conducting *in vivo* biochemical studies of the brain. Primarily a research tool but increasingly used in the courts, PET allows scientists to visualize and assess how various parts of the brain function under specific stimuli and to study the relationship between brain functioning and particular behavior. PET involves the injection of a radioactive material, such as glucose or fatty acids labeled with positron-emitting isotopes, into the body of an individual. This radioactive material decays by freeing a positron, which collides with an electron, thereby producing two photons that travel in opposite directions. A camera acts as a scanner to record what occurs when the two photons simultaneously reach detectors that are arranged opposite each other. The radioactive detectors record the activity level in various portions of the biological tissue under examination.[13]

PET is an expensive process, requiring an on-site cyclotron. At around $2,000 to $2,500 per test, its use for most clinical purposes is limited. However, it serves as a reference source for similar but far less costly technologies. SPECT (Single Proton Emission Computed Tomography), for example, functions similarly to PET but uses commercially available radioisotopes. While SPECT cannot provide quantitative information and has poorer spatial resolution, it is considerably less expensive (about $250 per test). This allows its utilization in the clinical setting. Thus, most of the nuclear medicine departments in the United States currently use SPECT as a routine diagnostic device.

A range of other sophisticated and less costly imaging technologies also address the functioning of brain tissue. Computerized Electroencephalography (CEEG) enhances the value of the traditional EEG because the computer allows for greater specificity in locating abnormal brain waves and facilitates the correlation of these brain wave patterns with external stimuli to establish the basis for detecting abnormalities.[14] Another version of the CEEG is Brain Electrical Activity Mapping (BEAM), which allows the electrical activity of a patient's brain to be compared with that of a healthy individual.

Another technology called SQUID (Superconducting Quantum Interference Devices) employs highly sensitive neuromagnetic sensors to detect magnetic fields that are generated by the brain. It measures these magnetic fields with such sensitivity that it can be used to visualize the electrical activity of neurons in the brain, detecting which parts are active under different conditions. By following electric currents, the process in effect "watches" the brain, observing how it functions. Computers then derive a three-dimensional image of what is going on.

Several new technological developments are expected to enhance the capacity to image biological processes. The development of synchrotron radiation—produced by the acceleration of electron beams to high energy—provides the opportunity to "tune" radiographs to certain elements such as phosphorous or magnesium and to trace bioactive chemicals within cells. Microcellular particles such as ribosomes are becoming the object of analysis. These technologies will allow the study of biological function at a highly focused and refined level.[15]

We are approaching a time when imaging techniques will be able to "see" how persons respond to sights and sounds, how muscles provide feedback for motor coordination, and how people pay attention. They will be able to monitor biologically the state of the brain when it is affected by sleep, drugs, nutrition, or the stimulation of ideas. And like genetic tests, imaging techniques can be used not only to detect the actual presence of disease in a population but to predict who may be predisposed to becoming ill at some future time.

Commercial Development

In 1987 the Office of Technology Assessment of the United States Congress (OTA) surveyed 120 biotechnology companies to assess the extent of the commercial development of diagnostic tests.[16] Of the companies that responded, 43 were working on products involving DNA sequences; 22 were developing or planning to develop probes for detecting genetic disease; and 8 anticipated that more and more people would have genetic profiles on record, and that genetic testing would become mandatory in many organizations.

As tests become available to identify common disorders with a genetic base, they are expected to be a part of routine medical care. Market forecasts, for example, anticipate that by 1992, about 30 million tests will be administered each year to predict susceptibility to common diseases with a genetic component. These include 12 million for cancer, 12 million for heart disease, and 5 million for diabetes mellitus. In addition, the market forecasters have predicted that 2½ million tests will be used annually for purely genetic diseases.[17] Considering that the total cost of such tests is estimated at about one billion dollars, it is understandable that market forces encourage their extended application.

Collaborative Research is one of the leading biotechnology companies in the field. Its founder and director, Orrie Friedman, says, "What we're working toward is genetic counselling for the common man."[18] He claims that genetic tests will reduce and possibly eliminate hereditary disease.[19] Collaborative Research was an early competitor in the race to develop genetic tests for hereditary diseases and, between 1984 and 1987, invested $11 million in probes for RFLPs. In its laboratories, investigators purify DNA from tissue samples and analyze them by using DNA probes to predict the inheritance of genetic diseases and to identify the presence or absence of markers. In November 1985 the company claimed to have identified the DNA marker linked with the cystic fibrosis gene, and a few months later offered a prenatal diagnostic test. It continued working on other single-gene disorders, later developing a test for adult polycystic kidney disease. In 1986 it obtained a license as a Diagnostic Reference Laboratory to offer clinicians and researchers a range of diagnostic tests associated with DNA markers. This laboratory provides, on a commercial basis, prospective analyses of risk factors for diseases with genetic components. Collaborative Research introduced the first gene map, with 403 markers, in October 1987. It is intended to facilitate diagnoses of common diseases involving multiple genes. Applying its technology to multifactorial common hereditary diseases, the company anticipates "a massive commercial market" in diagnostic kits.[20]

Another firm, Integrated Genetics, has developed six test kits for genetic disorders—cystic fibrosis, retinoblastoma, adult polycystic kidney disease, thalassemia, sickle-cell anemia, and hemophilia B. Well-established firms such as DuPont and Abbott Laboratories have negotiated deals to share their distribution systems and manufacturing facilities with biotechnology firms. In July 1986 Amoco Corporation signed a $20 million accord with Integrated Genetics to develop genetic probes jointly.

By 1986 a number of venture capitalists had begun investing in small neurobiology companies. An estimated fifteen companies were formed, each competing to develop the tools that would allow the diagnosis and treatment of neural disorders. According to Kevin Kinsella, an investor in Athena Neurosciences, venture capitalists are "falling all over themselves to get into the neuroscience field."[21]

In a related commercial venture, some companies are freezing or banking deposits of DNA until markers or genes become available. On request, the DNA bank will send a styrofoam box containing five vials to be filled with blood and mailed back with $50. Should the depositor have a harmful gene, he or she will be informed and guaranteed psychological counseling.[22] Such ventures have even become the subject of satire. A science writer describes a "vestibule of vials" where urine samples, blood samples, and skin tissue are stored in "Permanent Personnel Vials" ready to be analyzed for potential genetic problems.[23]

The rivalry among biotechnology firms working on diagnostic tests is intense, for the financial stakes are enormous. Cystic fibrosis, for example, affects 1 of every 2,000 live births, and 1 of every 20 Caucasians is a carrier. In 1988 Colorado proposed a law requiring that every newborn baby be screened for this disease. Its proponents see screening as a means to facilitate genetic counseling and informed choices for family planning. If screening becomes routine, the market will be worth hundreds of millions of dollars each year.[24] Thus Collaborative Research and other firms are engaged in fierce competition to find the gene and to patent diagnostic tests.

The market for DNA probes, used to identify genetic diseases, was $3 million in 1985. At that time some securities analysts expected it would be $4 billion by 1990. Such predictions were subsequently tempered. In 1986 *Genetic Engineering News* anticipated that the market value of tests for genetic and genetic predisposition to disease would come to $950 million to $1 billion by 1992. But at the same time, the director of the U.S. Office of Disease Prevention and Health Promotion predicted that most people would have genetic profiles on record by the year 2000.[25]

Professional as well as financial stakes encourage the extended application of new diagnostic technologies. Significant new technologies create their own professional and organizational configurations. Those fields of psychiatry that use imaging technologies as research tools are converging with neurology, a discipline increasingly concerned with behavior. The new profession of behavioral neurology now deals with those areas of medical diagnosis and treatment that were traditionally in the domain of psychiatry. Similarly, geneticists involved in DNA probe analysis have formed a new professional group identified by its ties to scientific and technological developments.

The availability of techniques that probe the relationship between behavior and the state of the brain has brought enthusiastic efforts to develop a new understanding of heredity and a "new psychological theory."[26] In the past, studies of the inheritance of behavioral characteristics relied on crude methods such as observation of identical twins or siblings who have been reared in different families. Such research has been used to explore the heritability of a wide range of characteristics: intelligence, personality, aggressive behavior, alcoholism, obesity, social attitudes, and even political conservatism. The emergence of biology-based tests encourages the belief that "just around the corner lies the vital new finding that will uncover the precise determination of unwanted and incomprehensible behaviors."[27]

But we must ask just how precise these tests can be. What are the problems of accuracy, reliability, validity? Diagnosis, after all, requires human judgment, and ultimately rests on a set of interpretive assumptions.

CHAPTER 3

Interpreting Tests

CONSIDER the dilemma of a twenty-five year old man who checks into the hospital for a neurological workup, including a Magnetic Resonance Imaging (MRI) scan. In the course of a routine examination, the MRI reveals an abnormality in the kidney, highly suggestive of polycystic kidney disease, an inherited, chronic condition. The patient has never experienced any symptoms of this condition and may never have them. Nonetheless, it could lead to total kidney failure. Moreover, if he has children they will have a 50 percent chance of inheriting the condition. They might develop symptoms by the age of twenty or they might live full and productive lives. How, then, is he to interpret the information from the test? Is the test sufficiently reliable to be trusted, and is the physician able to judge the signals appropriately? Assuming the condition has been reliably detected, will it ever be expressed in the disease?

Diagnosis is at the center of a contradiction between scientific medicine, involving research on statistical aggregates, and clinical medicine, involving the treatment of idiosyncratic individuals. Using diagnostic technologies developed from research in genetics and the neurosciences, physicians are detecting increasingly subtle abnormalities and predicting future impairments of physical or mental function. The statistical findings of these tests, processed by computers and imaged on a screen, appear to be objective, neutral, beyond refutation, equivalent to truth. Thus they assume considerable weight in drawing distinctions between normal and abnormal conditions—that is, in defining disease. But how is one to interpret such distinctions in order to predict? As a neurologist puts it: "The sensitivity of the tests is so high that they are picking up everything, and we're not really able to tell whether any of it has any bearing."[1]

Interpreting any test involves drawing on a set of assumptions about the accuracy and reliability of the instruments and the validity of the theories relating biological conditions to their expression in disease. These assumptions are valid in some situations and problematic in others. For example, in cases of physical pathology such as tumors or the presence of a metabolic disease, distinctions can be sharply defined. Patients who enter a physician's office because of physiological symptoms suggesting diabetes can quickly discover on the basis of a blood sugar test whether they have the condition, for the diagnostic test for blood sugar is highly correlated with the presence of the disease. However, distinctions are less clear in tests that seek to identify the root causes of physical or behavioral symptoms and to predict future conditions in the absence of manifest symptoms. In such cases correlation can easily be misperceived as causation, especially by non-scientists using tests for policy purposes. Such questions of interpretation become especially problematic as new technologies and the expectations they generate extend beyond the medical context to such institutions as the workplace, the schools, and the courts.

Interpretive Assumptions

The interpretation of data from diagnostic tests rests on several assumptions. The first is that the technology is *sensitive* to the presence of a biological condition allegedly existing in the tissue under examination. Second, it is assumed that the test is *specific*, able to distinguish meaningful from irrelevant conditions; that is, the marker of abnormality revealed by the test must be consistent with the actual behavior or pathological condition experienced by those who have an alleged disease. A further assumption is that the test is *valid*, that it is a more legitimate sign of disease than self-reported symptoms and therefore can be used to anticipate the presence of disease in the absence of overt manifestations.

These assumptions of sensitivity, specificity, and predictive validity underlie the interpretation of both genetic tests and imaging technologies. Both reveal markers for disease that are assumed to project a meaningful and valid description of a biological condition which can then be used to predict disease. The markers themselves are but "signs" of aberration, independent of any behavioral or physical dysfunction. Yet the results of such tests are granted more importance than symptomatology. It is assumed that the actual presence or future presence of disease will be revealed more accurately through the tests than through self-reported symptoms or behavior. Viewed as a more objective measure of abnormality, a test is often given precedence over other kinds of information available about the person being tested.

These assumptions become apparent in reviewing the diagnostic interpretation of PET. During a PET scan a person is given a positron-emitting compound in order to study which part of the brain is activated under different stimuli. The PET camera records the radioactivity in the person's brain and feeds this information into a computer. The computer creates a picture in an array of colors or shadings that correlate with the different intensities of the brain's metabolic activity as indicated by the level of radioactivity. The first part of the analysis, then, assumes the sensitivity of the instrument in detecting what is actually occurring in specific regions of the brain.

The assumption at the second level of analysis is that the statistically derived indicators of abnormality identify an actual behavioral aberration or pathological condition. Consider an experiment in which PET scans are taken of 100 individuals who are experiencing the same stimulus, for example, a violent television film. The patterns detected can be designated as indicators of a statistical norm. Suppose that 90 of the 100 subjects show symmetrical activation of similar sections of the temporal lobes of the brain. The other 10, however, reveal activation not only of the temporal lobes but also of the right frontal lobe. The spread of activation in these 10 subjects would be defined as statistically abnormal. And once this statistical pattern is established, the behavior of these persons is likely to be interpreted as supporting the abnormal description. Indeed, the "sign" of the PET scans could lead researchers to find behavioral abnormalities that would not otherwise be detected, and to interpret them as resulting from the brain's unusual activation pattern.

The third, and most significant, interpretive assumption will occur when, in a subtle shift, the PET image of an abnormal brain activation pattern begins to serve as a basis for predicting future behavior. Today it serves this purpose in the diagnosis of Huntington's disease. Diagnosis, especially in screening situations, can reveal unanticipated biological abnormalities. The very presence of a brain abnormality can then be defined as disease, though there may be minimal or no clinical or behavioral manifestations. Such test results, expected to indicate future problems, may lead to labeling a person (for example, as prone to disruptive or violent behavior) or to remedial intervention.

Interpretation of genetic tests is based on a similar set of assumptions. Genetic tests also produce "signs" of disease or future disease, but the complexity of genetic causation makes interpretation of these signs particularly difficult. While many human characteristics including height, weight, intelligence, longevity, and personality are generally recognized to be the product of both genetics and environment, this interactive model is not the one applied when an aberration or disease is discovered to be "genetic." Instead, as the understanding of specific genes becomes more precise, and normal and abnormal sequences of DNA are identified, clinicians tend to interpret all conditions known to have a genetic component as if genetics were the determining influence. This widespread tendency is partly based on projections from disorders such as Huntington's disease or cystic fibrosis, in which the presence of the gene is diagnostic. These single-gene defects have provided a relatively simple model that has become a paradigmatic example of how testing works. To apply the same kind of analysis to conditions such as obesity, alcoholism, or common disorders such as heart disease, however, is to overlook their complex etiology.

Valid and useful diagnoses, in other words, must not confuse the presence of a genetic or biological condition with the actual disease. Most genetic disorders, in fact, are polygenic, the product of the interaction of several genes with a person's environment. It is also important to realize that any given gene may be responsible for several traits. Even if a test can detect with complete reliability a gene, a cluster of genes, or an extra chromosome, it will not necessarily provide information about the timing or severity of a disability or how it might affect the normal functioning of the afflicted individual. Some diseases may be easily controlled through environmental changes. For example, phenylketonuria (PKU) is a severe inherited genetic disease that, under certain conditions, may never be expressed. Though sensitivity to phenylalanine is inherited, its principle manifestation, mental retardation, depends on diet. Removing phenylalanine from the diet of afflicted individuals will avoid the serious retardation that characterizes the disease. One can, in fact, have the gene, yet with proper dietary changes never show the manifestations of the disease.[2] In many other diseases as well, the presence of a genetic condition does not completely determine the person's ability to function. Environmental factors may intervene. Thus, increased accuracy in detecting the presence of a gene through the greater use of DNA probes will not eliminate the uncertainties in predicting future disability.

The linkage theories that seek to explain the genetic origin of neurological disorders, such as X-linked recessive manic depressive illness, demonstrate these uncertainties. Studies of family profiles had long indicated a genetic influence in affective disorders, but until 1981 no basic biochemical abnormality had been discovered. In that year a study of an Old Order Amish family found a specific locus for the abnormality on chromosome 11.[3] Using genetic segregation techniques, the study located two markers that had the same pattern of inheritance in all those family members who actually manifested symptoms of manic depression. While the markers are not the actual biochemical abnormality, they are located near the critical DNA on chromosome 11. Because they are the same in all family members with the illness, it is assumed that these markers relate to the presence of the specific genetic abnormality responsible for the manic-depressive disease. However, studies applied to other families failed to come up with the same genetic linkages as in the Amish, suggesting genetic heterogeneity; in other words, it is likely that more than one gene may predispose individuals to manic depression.[4]

At the same time, the presence of any one of these markers may not be predictive of manic-depressive illness in those with no symptoms. In the Amish study only 60 to 70 percent of those identified as carrying the gene for manic-depressive disorder actually became ill, and there were variations in when and how such illnesses were expressed. While technologies that allow detection of aberrant genetic structures enhance the accuracy of prediction, social or environmental factors can interfere with the biological transmission of traits.

Tests that identify genetic traits are intrinsically incapable of accounting for the other variables—diet, lifestyle, the effect of environmental or social interactions—that may influence their manifestation in disease. By choosing to focus on a single parameter in complex disease etiology, medical researchers minimize the possible role of these other variables. Thus genes become the "cause" of manic depression, even though 30 to 40 percent of those with the marker identified may not develop the disorder. Similarly, an aberrant brain pattern becomes the "cause" of a perceived behavioral aberration, even though no direct link between the two is possible from the data available through a PET test.

This scheme of analysis highlights the potential for diagnostic fallacies in tests that rely on inferential evidence. Assumptions that are statistically grounded may have little relevance to an individual case. More problematic, the very presence of a marker or a brain abnormality may be defined as disease when there are no clinical manifestations. For once a diagnostic procedure identifies the markers that are linked with diseases, clinicians tend to classify all those with the same markers in the same way. In time, routine use could obscure the uncertainties inherent in tests, their underlying assumptions could remain unquestioned, and the marker could become reified as the disease.

Reliability of Tests

Beyond the inherent problems of interpretation lie questions of reliability. A reliable test must produce a sign of disease that is highly correlated with both the self-reported symptoms and the somatic manifestations of a given condition, detectable, for example, through surgery or autopsy. The reliability of diagnostic tests has been under constant scrutiny, revealing persistent errors and contradictory interpretations. Early studies of the errors in reading X-rays, for example, found that misinterpretations stem from such factors as faults in materials, inadequate knowledge of the normal structural variance in the body, and the personal career biases of physicians.[5] Later studies of electrocardiogram tests revealed similar inconsistencies. Physicians reading the same EKGs agreed unanimously on only about one-third of the tracings. Reviewing the history of test reliability, Stanley Reiser, a historian of medicine and technology, found that "the crucial factor producing differences among observers was the lack of knowledge or agreement on the criteria for normalcy or pathology. . . . Precision in medical diagnosis seems to depend on three characteristics: the intrinsic accuracy of the measurement or test, the constancy of the phenomena being measured, the ability of the observer to interpret and record the phenomena."[6]

A recent study of autopsies at a university teaching hospital found that in 20 percent of cases, major diseases were discovered that physicians had either misdiagnosed or overlooked. In nearly half of these cases accurate tests might have led to effective treatment.[7] A subsequent study of autopsies and death certificates suggested the biases in clinical tests: circulatory and respiratory disorders were diagnosed far more frequently than they actually occurred, while other conditions such as gastrointestinal disorders and traumatic events were underdiagnosed.[8]

Problems of reliability are even greater in diagnosing psychological or behavioral conditions. Such conditions have been increasingly attributed to the structural or chemical abnormalities in the brains of afflicted individuals. The assumptions underlying this attribution have been questioned at least since 1941, when the work of D. Rothschild from Harvard Medical School suggested inconsistent relationships between behavior and brain structure. Using autopsies to examine the brains of those with diagnosed behavioral abnormalities, Rothschild found that the brains of some individuals with severe behavioral abnormalities had only minimal structural abnormalities. Conversely, some individuals with minor behavioral problems demonstrated major structural aberrations.[9] More recent studies have discovered frequent discrepancies between EEG images and the findings of pathological examinations. An individual may demonstrate severely abnormal behavioral patterns and yet an EEG will show only minimal organic brain damage. But EEGs sometimes find massive brain alterations in patients with minimal or no behavioral disorder.[10]

There has long been a tendency to attribute flaws in diagnostic testing either to problems with the technology or to human error on the part of technicians. The implication is that improved technology and automated systems that rely less on human judgment would create greater accuracy and precision. Certainly refinements in technology have improved the reliability of diagnostic testing. Yet even the most sophisticated medical technologies produce test results that require interpretation by human operators who are inevitably influenced by a set of preconceptions and biases.

Psychologists Amos Tversky and Daniel Kahneman have studied the systematic character of interpretive bias. They have observed that even experts frequently ignore objective information and make important choices based on bias-laden heuristic reasoning.[11] They found, for example, that experts tend to link events on the basis of their resemblance, and are frequently insensitive to the uncertainties revealed by previous studies. They also found that experts base their notions about the probability of an event on the ease with which an occurrence of a similar nature can be brought to mind. And the easy availability of explanations, from previous studies or from personal experience, contributes to the biased approach to new material. A school physician confronted with a student who cannot sit still may on the basis of experience diagnose the condition as attention deficit disorder or minimal brain damage. In arriving at this diagnosis, he may ignore a wide range of other conditions that can produce the same symptom: certain forms of rheumatic fever, a serious emotional disruption such as the death of a parent, or a viral infection.

Other studies of clinical judgment emphasize that experts tend to pursue supportive evidence selectively, undervaluing counter-evidence. Typically expecting to find abnormalities, they can usually "succeed" in finding them.[12] This bias frequently leads clinicians to overpathologize—to identify abnormalities when this is inappropriate. While improved technology and the use of sophisticated statistical methods and computer programs may minimize operator judgment, such interpretive biases will persist and continue to influence policy and practice.

As technologies improve, allowing the detection of even minor aberrations that might indicate the presence of a disease, judgmental biases may have critical social implications. For example, refinement of the sensitivity of diagnostic technologies can encourage misleading interpretations that may be socially damaging. An absurd example is the drug test on a paroled prisoner that detected "heroin" after he had eaten a poppy-seed bagel. More common are test results that detect very small deviations, such as a minor amount of blood in urine or slightly high blood pressure. While these may have minimal consequences for health, they can be used to exclude a person from insurance or employment.

While all tests must identify the presence or absence of a pathology with reasonable reliability, the level of "reasonable" accuracy will differ according to the context in which the test is employed. In a clinical context a test that identifies a problem that does not exist (a false positive) is generally less damaging than a test that fails to identify an existing condition (a false negative). If a test picks up a condition that does not exist, subsequent tests can correct the error. But if the test does not pick up an existing condition, for example, an early cancer, the individual may be lost to the health care system at a critical moment in the evolution of the disease. In nonmedical contexts, however, false positives may be more damaging than false negatives. The reliability level necessary for a given test, in other words, depends on the ramifications of the test results.

Screening programs involving large numbers of people are intrinsically less precise than clinical diagnoses and yield a higher percentage of false results than would be acceptable in the clinical context. The false positive rate of any test is directly related to the prevalence of the condition tested: the rarer the condition, the higher the false positive rate. Suppose a test is 100 percent sensitive (that is, it discriminates every person with the abnormality) and 90 percent specific (meaning 10 percent of those studied will be mislabeled). If the prevalence of the condition is 1 percent of the whole population, as in schizophrenia, then if everyone is tested, ten false positives will show up for every true positive. If the prevalence is 0.1 percent, the number of false positives will increase tenfold. But if just high-risk groups are tested, of course, the rate of false positives will decline.[13]

For the institution carrying out the screening test, a low level of reliability may be adequate to meet its needs, for example, for long-term planning or for the allocation of resources. But for the individual being screened, errors may have very high costs. A diagnosis that falsely labels a school child as hyperactive can have distressing consequences, and it may take a long time to discover the inaccuracy of the test. A false positive diagnosis of AIDS would be devastating, as persons falsely diagnosed as seropositive experience not only emotional and psychological distress but also the social stigma attached to the disease.

Despite the limits of screening as a guide to individual pathology and the consequences of unreliable diagnosis, test results are increasingly perceived as accurate indicators of disease and continue to be used to justify social action.

Beyond the Clinical Context

The problems of diagnostic uncertainty are greater when a technology developed to identify a pathology in the context of clinical care is transferred to another institutional setting for use as a screening tool. This is frequently the case when tests that were developed to identify the behavioral or psychological problems of patients are used for screening in the workplace or in other nonmedical settings—in the institutions that oversee health care financing, education, incarceration, prosecution, or work. Genetic tests for sickle-cell anemia, for example, were developed to help carriers of the disease make informed family planning choices. But the tests have been used in the workplace to identify those susceptible to illness from exposure to chemicals and to exclude them from employment opportunities. Analyses of genetic linkages, intended for use as a basis for genetic counseling, can also be used to identify those who may be too great a risk for an insurance plan.

In a clinical situation, inconsistencies are easily discerned because the purpose of a test is to discover the abnormalities underlying a single individual's symptoms. But when tests are used for screening purposes, where the objective is to deduce statistical levels of disease in a large population, individual inconsistencies can remain undetected, and the potential for misdiagnosis, with all its problematic consequences, is far greater.

With the transfer of clinical tests, the subtleties of interpretation may be lost as the objectives and values of the institution using the test results shape the definition of abnormality. For example, a diagnostic test in the clinical context may reveal an aberration that would be interpreted within a range of normality; the same test used to evaluate eligibility for an HMO may suggest a reason for exclusion. For the very pressures for efficiency and cost control that encourage efforts to predict the potential diseases of a client population can overshadow the uncertainties of screening techniques.

As new diagnostic technologies are able to detect a wider range of problems with increased sensitivity, they are also more likely to be used in nonmedical contexts. Take, for example, the case of alcoholism. For many years the research on alcoholism has been based on assumptions that the disease runs in families. Studies comparing the history of siblings raised in different families found that those with alcoholic fathers were likely to become alcoholic themselves. Today research on alcoholics is increasingly carried out at the molecular level. Scientists are seeking the genetic traits that predispose certain people to alcoholism. For example, some studies have found that certain biological abnormalities in platelet monoamine oxidase activity resulting from alcohol abuse occur only in people who are supposedly genetically predisposed to alcoholism.[14] In addition, brain-wave studies reveal that nondrinking children of alcoholic parents show characteristic neurophysiological abnormalities, and these are interpreted as measures of an underlying genetic susceptibility. These abnormalities indicate the possibility of developing biological tests that would allow diagnosis of a predisposition to alcoholism well before either the patient or the physician became aware of its presence. Indeed, according to a scientist working in this area, it is possible that "persons at risk for alcoholism could be identified before they began drinking."[15]

The evidence supporting such prediction is fragile—and many of those with metabolic abnormalities may never become alcoholics—yet the possibility of prediction is seductive. Alcoholism is a costly disease. In 1980 it was estimated to cost at least $90 billion, including accidents, medical care, and reduced productivity.* As the potential for predictive screening develops, there are powerful economic incentives to use it, even when evidence is weak.

Schools, employers, and the courts all stand to gain from better understanding of the present and future health status and behavioral syndromes of their clients. In these contexts, diagnostic technologies are a means to facilitate planning and reduce costs. In the health care system, DNA probes and imaging technologies can provide the basis for profiling and predicting the long-term diseases and behavioral characteristics of prospective patients. They can provide predictive parameters for insurance companies designing premiums, for schools attempting to assess the potential of students before admitting them to lengthy and costly special educational programs, for health maintenance organizations seeking to anticipate the possible development of disease among their clients—indeed, for any organization concerned with problems that might contribute to future costs. With all their uncertainties, therefore, long-term health predictions and biological profiles are likely to be added to the increasing amounts of information already set into the charts and computer programs of patients, students, and employees. And the institutions controlling this information can impose their values by tailoring it to meet their needs.

*It has been suggested that this is an underestimate since alcohol abuse is difficult to diagnose and often omitted from hospital records.[16]

CHAPTER 4

Diagnosis in the Health Care System

IN OCTOBER 1986, the Society of Actuaries met to discuss the future of underwriting in light of changes in risk-classification practices. Insurance companies had greatly increased their requirements for information about the health status of applicants and had added new classifications designed to attract a "preferred" group of customers who could be expected to incur fewer medical expenses and therefore pay lower rates. As the vice president of Prudential Insurance Company explained: "We now see the development of super-preferred classifications for individuals who don't smoke, have wonderfully low blood pressure, no medical histories and white collar occupations, and who are able to swear to the fact that they exercise three times a week at Jack LaLanne's." His own company increased its classification categories from ten in 1980 to nineteen in 1986 and was contemplating the creation of still more risk groups. He expects to make increasingly subtle distinctions on the basis of biological tests. "We undoubtedly will see improvements in these techniques. We've even seen a series of tests claimed to accurately predict an individual's biological age [life expectancy]."[1]

Prediction on the basis of biological testing meets administrative needs of insurers struggling to control medical expenditures. It also helps hospitals, individual physicians, and health maintenance organizations rationalize their administrative planning. The need for cost containment has joined fear of litigation to intensify the search for more "objective" indicators of patient status and more predictive diagnostic information to guide day-to-day management, legal self-defense, and long-term planning. Patient management depends on the efficient allocation of resources, and diagnostic information can enhance efficiency. Protection against malpractice depends on the legitimacy or defensibility of diagnostic decisions. And long-term planning requires an ability to predict future health care demands. Thus diagnostic technologies are of growing importance in guiding not only patient therapy but also administrative strategies throughout the health care system.

The adoption of new diagnostic techniques that can predict late-onset genetic diseases and even common behavioral disorders will reflect the strategies of control currently used by health care providers to reduce costs, to minimize the risk of malpractice suits, and to plan for the efficient use of health care resources. Indeed, a striking aspect of new technological developments lies in their potential use by health care providers to serve their own economic and administrative purposes.

Pressures on the Health Care Delivery System

The health care delivery system has changed dramatically since the early 1970s in ways that have encouraged strategies to predict and control patient populations. The rapid evolution of for-profit corporate hospital chains has challenged fundamental assumptions about the economic organization of health services.[2] The growth of prepaid health delivery plans has reduced professional autonomy and encouraged attention to issues of profit and loss in everyday treatment decisions. The threat of malpractice litigation has created powerful incentives to justify and legitimize health care decisions with objective technical evidence. Government reimbursement plans linked to specific diagnoses and the demands on private insurance have called for greater institutional efficiency and cost control. And consumer demands in a competitive medical marketplace have encouraged a growing use of diagnostic technologies as people come to expect sophisticated tests as a standard part of quality medical care.

These pressures force institutions to search for "profitable" patients. Prepaid medical plans seek patients who are likely to have relatively few costly illnesses. Hospitals seek patients who can offer adequate third-party reimbursement for their medical care. Third-party insurers are requiring more and more information about the health status of applicants who are not on group plans.* Diagnostic tests can facilitate an institution's ability to plan for future economic imperatives, and they can provide the technical justification for potentially controversial health care decisions.

THE CHANGING STRUCTURE OF HEALTH CARE DELIVERY

In 1970 Paul M. Ellwood, Jr., a Minneapolis physician, proposed the concept of "health maintenance" through prepaid group practice. Spurred by both patients' rights advocates and bureaucratic promoters of efficiency, the Nixon administration embraced the idea of the health maintenance organization (HMO), and by 1987 nearly 30 million people were enrolled in 663 prepaid health plans across the nation.[4]

The economic priorities of prepaid health financing arrangements and the pressure from employers who pay most of the bills

*About 14.5 million non-Medicare individuals and their families have health insurance without the benefit of group membership.[3]

have encouraged rigorous cost controls.[5] Those managing HMOs have a stake in predicting and preventing illness and in controlling the use of medical facilities. Competing for clients, they must offer an attractive range of services, but they must also limit day-to-day operating expenses and accurately predict future costs. Critics of HMOs view them as a system of medical rationing disguised as cost containment.[6] To hold down costs, some HMOs require the doctors working in the hospitals to sign fixed-priced contracts where they assume the economic risks if they provide services exceeding the contractual agreement. One means of reducing both day-to-day and long-term costs is to control the patient population; thus a careful screening process precedes HMO membership.

The rise of multihospital systems is also shaping diagnostic practices. Independent hospitals began to give way to multihospital chains in the early 1960s. By 1978 there were 155 multihospital systems, so-called super-meds, in operation in the United States; 34 of them were investor-owned, for-profit chains, and 121 were nonprofit chains run by religious groups or charity organizations.[7] The for-profit chains, fueled by private investment and run with special attention to reaching high-income consumers grew rapidly, to control about 12 percent of U.S. hospitals by 1986.*

The super-meds, both public and private, benefit from economies of scale and other planning advantages of large corporations. An executive from Humana, a major for-profit hospital chain, compares the goals of these organizations to the marketing goals of McDonald's: uniform standards across the country.[9] And like McDonald's, the super-meds value efficiency and profitability. Through careful attention to the characteristics of their patient population, the for-profit chains work to limit economic uncertainty.

*Humana, Inc., for example, began in Louisville in 1968 with a few nursing homes and $4.8 million in revenues. By 1984, it had ninety-one hospitals and nearly $2 billion in revenues.[8]

As the health care industry has come to be dominated by fixed prices for medical reimbursement, it has focused a shrewd eye on the patient population. In 1980 the federal government introduced the concept of diagnostic related groups (DRGs) as a guide to Medicare reimbursement of health care providers. Hospitals must classify patients with similar clinical conditions into DRG categories, with specific reimbursements designated for each diagnostic group. DRG rates take into account an individual hospital's cost for providing care to specific types of patients and the statewide average cost for all hospitals. By late 1983 DRGs became the standard for Medicare and Medicaid payments, and some private insurers began to adopt the concept of the DRG to guide their own reimbursement rates. Hospitals either retain the difference between the reimbursement rate and the actual cost of providing care or assume the loss if the cost of care exceeds the amount of reimbursement.[10]

The private insurance industry also responds to economic pressures by minimizing risks through selective underwriting. This requires the industry to identify and classify high-risk applicants. People under the age of sixty-five who are not on group policies must be individually evaluated for their eligibility for coverage. Underwriting procedures usually require a medical history, including data on family illnesses, a statement, and, in some cases, records from an attending physician. Occasionally special tests are requested.

These economic pressures have upset traditional relationships within the hospital, increasing administrative control and reducing the power of the physician, whose work is evaluated partly on the basis of cost effectiveness. Some hospitals revoke physician privileges if individual cost profiles exceed DRG reimbursements; some reward those physicians who keep costs down. There is in this system considerable incentive for diagnostic manipulation and patient selection on the basis of anticipated cost.[11]

And though physicians tend to deny it, studies suggest that diagnostic decisions are often shaped by economic motives.*

LITIGATION

In 1983 the FDA approved the use of alpha-fetoprotein screening to detect neural tube defects in the fetus. However, the American College of Obstetrics and Gynecology opposed unregulated use of the test, urging that it should be implemented only when combined with a coordinated system of care and follow-through services. Only two years later, ACOG's Department of Professional Liability advised its members that "it is imperative that every prenatal patient be advised of the availability of this test and that your discussion about the test be documented in the patient's chart."[13]

The threat of litigation drives the practice of testing in the health care system. Physicians are inclined to offer tests as soon as they are available, for litigation has established the physician's obligation to warn patients of potential problems if it is possible to gather such information through diagnostic technologies. In *Curlender v. BioSciences Laboratories* (1980) a Tay-Sachs victim, Shanna Curlender, sued a laboratory on a "wrongful life" cause of action for providing inadequate information about her parents' status as carriers of this genetic disease. The California Court of Appeals cited the "dramatic increase in the last few decades of the medical knowledge and skill needed to avoid genetic disaster."[14] In light of this knowledge, it ruled that the laboratory had a duty to provide accurate information, which would give parents the choice to prevent the birth of children with serious genetic illness.

In another example, *Schroeder v. Perkel* (1980), parents sued for the "wrongful birth" of their second child who, like the first, was afflicted with cystic fibrosis, a fatal genetic disorder. They claimed that the pediatricians had failed to make a timely diagno-

*See, in particular, the increasing use of endoscopy, an expensive technology that has virtually no effect on treatment but a dramatic effect on physician income.[12]

sis of cystic fibrosis in the first child. As a result the parents were not warned about the possible presence of the condition in the second child until the eighth month of gestation, when an abortion was no longer possible. The New Jersey Supreme Court awarded the parents extraordinary medical expenses for raising the second child.[15]

By recognizing such "wrongful life" and "wrongful birth" claims and by awarding damages to parents of children born with a serious genetic disorder that could have been identified prior to birth, the courts have implied that failure to use existing screening technologies constitutes breach of duty. In other words, the availability of diagnostic tests implies the obligation to use them, or at the very least, to advise patients of their availability.

The threat of malpractice has become a critical variable in shaping daily operations, long-term planning, and patient treatment. In numerous discussions, physicians and medical administrators present malpractice litigation as a "serious dilemma," a "serious disease," and a "major threat" to institutional integrity that requires constant vigilance in the handling of patients. They portray the modern hospital as a besieged institution aggressively engaged in self-defense. Hospitals, for example, are responsible for counting clamps after surgery—crucial in the avoidance of "preventable risks," such as leaving a foreign body inside a patient. Such a mistake, immediately recognizable to a jury, does not depend on controversial definitions of "quality care," and usually results in large damage awards to the plaintiff.[16]

The cost of malpractice insurance has caused some physicians to abandon certain types of surgery. A 1984 study found that one surgeon in four had stopped performing some operations in response to the threat of litigation. But fear of malpractice suits has especially influenced the way practicing physicians make decisions about patient care. According to the AMA, "defensive medicine" involving overuse of diagnostic tests is costing about $15.5 billion a year. However, this estimate was backed by inadequate studies and does not make clear how many diagnoses were harmful to the patient.[17]

Existing Strategies to Meet Institutional Needs

Economic and legal pressures converge on health care institutions to encourage strategies of self-protection. One important strategy, commonly known as "skimming," is the effort to allow access to insurance and medical care facilities only to those who are expected to be "profitable" patients. Another is the use of standardized planning techniques that facilitate optimal use of hospital services. Still another strategy is the use of consulting psychiatrists to mediate controversial decisions. These strategies all rely on increasingly refined diagnostic technologies.

RESTRICTING ACCESS

Restricting access to medical care is a well-established cost-control strategy. Hospitals often close their doors to so-called 066s, those patients without medical insurance or other means of paying.[18] These incidents are merely blatant examples of what other health care providers do with more subtlety.

In 1984 a medical economist described the pervasive shift in the relationship between medical providers and their consumers as follows: "The name of the game is skimming and this is no longer frowned upon. All providers are involved in this process and only those who do it best will survive. Skimming will become an art deeply impregnated with the highest cultural value of success."[19]

A great many studies in the 1980s have pointed to the "transformation" of medical care, suggesting that increased price competition, privatization, and commercialization have reduced the access to quality care for "unprofitable" patients. These may include patients who are uninsured or who have chronic conditions that take up the time of the physician but require few reimbursable procedures. For an individual physician, the repercussions of continuing to care for such patients can be serious. One endocrinologist, firmly convinced that appropriate care of her diabetic patients required careful, time-consuming attention to their dietary regimens, found herself in debt and forced out of practice because such care, which involved no procedures, was not reimbursable.[20] Such incidents reinforce professional concern about the economics of treating patients with certain chronic conditions.

Predictive information about prospective patients can enhance economic efficiency. Although a hospital emergency room can arrange to transfer those unable to pay, keeping problem patients out of an HMO or an insurance pool depends on being able to recognize health problems early. Genetic screening technologies that predict predisposition to a disease prior to the onset of symptoms provide precisely the kind of information needed by health care providers.

Insurers and health maintenance organizations involved in health care financing are keenly interested in the characteristics of the populations they serve. Risk-selection tactics in the health insurance industry include entry barriers and differential rates. It is standard practice in the insurance industry to refuse to cover people with anticipated serious health problems and to place limitations on reimbursement for existing conditions. In 1987, 20 percent of insurance applicants were classified as substandard; their policies excluded particular conditions, or they paid higher than usual rates. One company, for example, would not give preferred insurance rates to a person whose parents contracted cardiovascular disease before the age of sixty. In addition, 8 percent of applicants were judged uninsurable.[21] Some genetic conditions are considered unacceptable for either medical coverage or disability insurance. These include autism, spina bifida, duodenal or gastric ulcer, narcolepsy, and active rheumatoid arthritis.[22]

HMOs and other prepaid programs also limit membership on the basis of anticipated health. In 1987, 24 percent of the individual applicants to HMOs were denied membership. HMOs require applicants to have thorough medical exams to detect conditions that might in the future develop into a costly illness. Indication of high blood pressure, for example, can exclude a person from an HMO because it suggests borderline hypertension. New technologies, detecting even more subtle conditions, allow these organizations to further refine their eligibility requirements so as to lower their costs.

For-profit hospital chains have also controlled the characteristics of the populations they serve. They built new hospitals in fast-growing Sunbelt states where regulation is less restrictive and in areas with lower Medicaid and indigent patient loads. Humana, Inc., for example, builds primarily in suburbs full of privately insured young people and tailors its services to the needs of the middle class. A 1986 survey of physicians found that 40 percent of those in investor-owned systems in competitive markets believed their institutions were restricting access, while only 16 percent of those in nonprofit, noncompetitive environments reported such restrictions.[23]

Diagnostic and screening technologies that provide detailed patient profiles or predictive information readily conform to existing management strategies emphasizing patient selection and exclusion. If mass screening revealed the presence of a high rate of future disease in a certain region, this could be an important variable in management decisions about the provision of future services. If diagnostic technologies identify and predict the future health problems of potential clients, those whose problems might tax the economics of an institution could be excluded.

EFFICIENCY TECHNIQUES: ALGORITHMS

The process of diagnosis, once dependent on insight, observation, and personal judgment, increasingly relies on diagnostic tests that minimize individual decision making. Technology is considered a more efficient way to decipher patient symptoms, and health care providers are turning to standardized testing sequences that minimize individual discretion. The use of algorithms and decision trees illustrates the tendency to seek more scientific grounds for diagnostic decisions through the systematic use of tests. Algorithms help to categorize and channel patients according to statistical probabilities related to their complaints. Those complaining of certain pains, for example, are given predetermined tests in a sequence that is statistically most likely to reveal relevant information. The sequence of tests is based on the information already known about a condition, the importance of its detection, the penalty for delay, and the least time, risk, and inconvenience for the patient.[24]

Clinicians also use decision trees to help them in patient prognosis. Decision trees derive from operations research and game theory. The tree is used to represent the strategies available to the physician and to calculate the likelihood of specific outcomes if a particular strategy is employed. The relative value of each outcome is described numerically, and a statistical conception of normality helps define proper patient care.[25]

Such techniques can help to avoid premature closure of a line of investigation. This may be crucial to proper diagnosis, especially in cases involving a rare disease. Decision analysis forces the physician to consider all pertinent possibilities. However, algorithms and decision trees can be arbitrary and inapplicable when symptoms are highly complex. In effect, they reduce the judgment of physicians to the interpretation of mathematical probabilities. Some critics have argued that informed human judgment is more akin to constructing a map than to following an algorithmic sequence. Unlike a map, an algorithm is linear and selective; it does not permit changes in the order in which information is found, for the sequence is intrinsic to the process.[26] Yet such flexibility is necessary in diagnosing the complex and interconnected variables relevant to understanding disease. Despite continued ambivalence about the value of algorithms and related computer-driven sequences, their use expands, reflecting the priority of efficiency that guides diagnostic decisions.

USING CONSULTING PSYCHIATRISTS

Hospital administrations sometimes require decisions that involve judgments about the competence and mental state of patients. Often they employ consulting psychiatrists as a strategy to meet their institutional needs. Is a person sufficiently competent to refuse the treatment recommended by the physician? Is a patient able to decide whether to be transferred from a hospital to a nursing home? Is a difficult or recalcitrant patient "crazy" or merely angry or uncooperative? These decisions, often controversial, involve a complex mix of moral and medical considerations. Diagnostic technologies can be used to minimize moral dilemmas by providing concrete biological evidence of impairment.

Consulting psychiatrists are frequently employed to resolve such disputes. They use various diagnostic procedures to assess whether patients are sufficiently competent to assume control over their own care. Psychiatric assessments have in the past been based on phenomenological evidence—derived essentially from interviews and the history of the patient's behavior. The introduction of new biological technologies such as computer-assisted EEGs, SPECT, and neuropsychological examinations is enhancing the importance of the psychiatrist's role in these complex decisions.

For example, a physician in an intensive care unit, caring for a cardiac-impaired patient who refuses to allow a catheter to remain in his arm, could justify coercing treatment if a psychiatrist, using a computerized electroencephalograph, demonstrated abnormal activity in the temporal lobe. This would suggest that the patient is suffering from a dissociative reaction or psychotic episode and is consequently incompetent to understand the nature of his actions.

Improvements in diagnostic capability are expanding the role of the consulting psychiatrist. Fortified by biological information, the psychiatrist is increasingly serving as a mediator, or judge.[27] The result has been to minimize the social and moral questions involved in such decisions by redefining the issues in biological terms.

The Use of New Diagnostic Technologies

Health care consumers expect and demand sophisticated diagnostic services as a standard part of quality medical care. Their demands are often based on the misleading assumption that diagnosis will logically and inevitably lead to effective treatment and cure, expectations that reflect, in part, mass media coverage of new medical technologies: the "promising new discovery," the "magic bullet."[28] But consumer demands also follow the commercial promotion of new technologies, which are frequently considered symbols of prestige in an increasingly competitive market.

The diffusion of new technologies is a complex process. In part, the market is driven by third-party payers such as Medicare, for they decide whether a technological procedure is reimbursable. But the decisions of third-party payers depend on physician judgment about clinically acceptable techniques. Physicians, in turn, are influenced by the incentives involved in using costly technologies. This process can push a technology like PET into clinical acceptability even before the scientific advantages are clearly established.

Financial incentives to promote technology are amplified by the competition to attract patients. Clinics often hold press conferences to announce the acquisition of new technology, as a Paramus, New Jersey, clinic did in July 1988.[29] Announcing a new optical spectroscopy machine, the president of the clinic claimed the technology could indicate whether a woman will develop breast cancer, quantifying her risk level from 0 to 10. The American Cancer Society does not condone optical spectroscopy as a screening technique, but clinic administrators were counting on the tremendous appeal of this new technology, and the fear of cancer, to attract customers.

Certainly customers have begun to demand the latest technology. The rapid rise in the demand for prenatal testing is an important case in point. Encouraged by the promotional efforts of biotechnology firms, fear of birth defects, and subtle pressures from genetic counselors and other health professionals, prospective parents have embraced amniocentesis, chorionic villus sampling, and other diagnostic technologies that can predict the outcome of a pregnancy.

GENETIC TESTING AND COUNSELING

A large and growing number of genetic conditions can be detected through fetal screening techniques, which can identify genetic abnormalities in the placenta when women are only eight to ten weeks pregnant. The alpha-fetoprotein (AFP) test, done at fifteen weeks, is a blood analysis used to detect neural tube defects. An abnormal amount of AFP being released by the liver can signal other defects as well. Over 180 fetal disorders can be detected through such tests.[30] These range from serious conditions, such as Tay-Sachs disease, Down's syndrome, and Duchenne muscular dystrophy, to neural tube disorders to relatively mild impairments, such as cleft palates. Based on current indications for amniocentesis, less than 5 percent of detected fetal abnormalities are considered serious enough to warrant consideration of abortion.[31]

Prenatal testing is now a routine part of obstetric services for women over thirty-five years of age. In many cases, it is also available to younger women identified as subject to "maternal anxiety." In the case of the most devastating conditions, prediction through prenatal tests is clearly desirable, opposed only by some right-to-life groups. But in some cases, demands for genetic services reflect a desire to control reproduction in more questionable ways. Critics believe that new technologies have created new standards of perfection and a desire among women for an "improved product."[32] More and more genetic counselors encounter couples who want prenatal tests only to determine the sex of the fetus. A survey in the United States found that 32 percent of 295 geneticists were willing to test for purposes of sex selection.[33] The interest in sex selection is evident in the proliferation of clinics using new techniques that can separate the X from the Y chromosome. The sponsoring firm, Gamete Limited, advertises a 75 to 80 percent success rate in determining the sex of the child. Another company, Gender Choice, markets home ovulation prediction kits (in pink and blue) as a way to preselect a child's sex.

In cultures where the sex of offspring has serious economic ramifications, such as in India, the use of amniocentesis is already a well-recognized problem. According to a *New York Times* report, tens of thousands of female fetuses are aborted each year.[34] The experience in India suggests that if incentives exist for such manipulation, it is likely to occur, and in American culture powerful incentives for the "perfect child" do exist. This social emphasis creates a climate in which attributes such as a slight diminishment in intelligence, a statistical likelihood of mental illness, a susceptibility to alcoholism, the possibility of heart disease in middle age, an anticipated birthmark, or even short stature could become the basis for abortion. Genetic tests create such possibilities, so it can be expected that some people will exploit them.

As the number of diseases defined as biogenetic has expanded, so too has the role of genetic counseling services. For example, despite its complex origins, mental illness is increasingly defined in biological terms. In 1988 the government funded the University of Maryland to set up a genetic counseling program specifically focused on mental illnesses, explicitly accepting the definition of such diseases as biologically inherited.

For most serious or chronic genetic diseases, genetic counseling offers the possibility of aborting fetuses with disabling or potentially disabling conditions. Thus, genetic information is of enormous value to prospective parents who might be faced with decades of anguish and ruinous medical costs. But health care providers also have an economic stake in favoring abortion: it avoids the future expenses of patient care. Children with a disabling condition may require a lifetime of expensive health care. A number of studies have documented the cost of genetic disorders, believing them to occur in 3 to 5 percent of all live births and to account for 20 to 30 percent of all pediatric hospital admissions and 12 percent of all adult hospital admissions in the United States.[35] Genetic services thus accommodate institutional imperatives as well as consumer demands.

The presumed beneficiary of genetic counseling is the prospective parent. But health care providers and also insurers benefit from genetic information; for reproductive decisions may have consequences for future medical obligations. An HMO that may have to finance the care of a handicapped child, for example, has an intrinsic interest in influencing the reproductive decisions of individuals in ways that will minimize future burdens.

From time to time proposals are made to reduce the costly burden of genetic illness. Some have eugenic overtones. In 1976 the Supreme Court of North Carolina declared constitutional a statute authorizing sterilization of the mentally ill on grounds that "the State has a right to prevent the procreation of children who will become a burden to the State."[36] Similarly, the Chicago Bar Association has argued that Illinois marriage laws should require premarital tests for diseases causing birth defects and has suggested that the State require correction of genes for specific maladies when such possibilities become technically feasible.[37]

Several states have chromosome registries which, for research purposes, keep track of the number and type of genetic defects identified through cytogenetic diagnosis. In 1981 the New York State Birth Defects Institute requested that all registered laboratories undertaking cytogenetic diagnosis supply not only the number of cases but also the names, addresses, and other identifying data about all individuals with chromosomal abnormalities. Laboratories providing genetic services would have to obtain permission from those undergoing tests to disclose such information to the State. Concerned about issues of confidentiality, a group of genetic laboratories refused to comply and the State dropped the request. However, in 1987 the Birth Defects Institute renewed the request, imposing additional information requirements, including the patient's social security number. The stated purposes of the 1987 request were to facilitate research that could identify environmental and biological factors possibly associated with fetal abnormalities and also to improve the quality of medical care. Again, several laboratories refused to comply and the requirement was suspended.[38]

The norms that guide genetic counseling programs specifically emphasize nondirective advice. Their very purpose is to expand the parents' choice over conditions that were once ascribed to fate and viewed as beyond control. But alternatives in fact may be limited, as screening technologies that predict genetic illness turn fated conditions into voluntary decisions. It may appear that the choice to risk the predicted situation is a voluntary one, and voluntary choices are traditionally treated differently from those that are involuntary. In particular, society presumes that the financial burden for voluntary decisions legitimately belongs to the individual making the choice. Take, for example, a mother who chooses to bear a child with a serious and costly illness, predicted by an amniocentesis early in her pregnancy. Could her decision in the face of warnings have implications for deciding who should bear the cost of the necessary medical interventions?

Despite their goal of nondirective genetic counseling, clinics do differ in their results, suggesting the variation in their approaches. In a study of the reproductive patterns of families who have a child born with phenylketonuria (PKU), researchers found that genetic counseling clinics reported very different responses among their clients. In some clinics as many as half of the families counseled went on to have more children. In others, none of the counseled families had further births. The study concluded: "This variation raises questions regarding the extent of direction that actually takes place in reproductive counseling, and the underlying philosophies of clinicians regarding families' needs to modify their reproduction plans when PKU is identified."[39] It suggests that even nondirective counseling can influence decision making and subtly shape future populations. This has not been ignored by right-to-life groups, who are strongly opposed to genetic screening and counseling programs, fearing that the development of diagnostic technologies is encouraging abortion. They believe that the focus should be on fetal rights rather than on the parental right to know.[40]

PSYCHIATRIC EVALUATIONS AND THE DEFINITION OF "NORMAL BEHAVIOR"

Research employing new diagnostic capabilities has contributed to refinements in the classification of disease. In turn these classifications have encouraged the expanding use of tests in medical contexts, especially for mental illness or behavioral problems. Through their ability to detect more and more subtle deviations from the norm, new tests have increased the number of diagnostic labels used to describe disease. These refinements in diagnostic labeling have been especially evident in the changing definitions of "normal" and "abnormal" behavior.

Traditionally the discipline of psychiatry had assumed a holistic approach to patients and emphasized the symptoms of mental disease as opposed to the classification of disease conditions. Though using medications to modify symptoms, psychiatrists have relied heavily on psychotherapy as a means of resolving the psychological conflicts underlying emotional problems. In the last twenty-five years, however, the profession, seeking to reintegrate itself into medicine, has looked increasingly to biological causation and cure. Influenced by data provided by neurological tests, the medical and scientific community is developing a consensus that diseases involving human behavior originate in biological processes. The latest revisions of the American Psychiatric Association's diagnostic manual of behavioral abnormalities (the 1980 *Diagnostic and Statistical Manual III*, or *DSM III*, and the 1987 *DSM IIIR*), for example, reflect these new assumptions. Though avoiding discussion of the etiology of disease, the *DSMs* have focused on categorizing disease and on refining and expanding the classification of disease conditions.[41] Indeed, the 1987 edition included twenty-five disorders that were not in the previous version.*

These classifications are important. The *DSM* is a widely used reference, providing standardized diagnostic criteria not only for clinical practice but for psychologists in schools, lawyers in the courts, and social workers; that is, wherever judgments must be made about mental health. Moreover, its increased emphasis on disease classification places priority on developing techniques of differential diagnosis, inviting the growing use of medical tests such as PET, EEGs, and the development of genetic markers to better understand and diagnose the biological roots of mental ill-

*These recent trends in differential diagnosis are in keeping with the Kraepelinian approach to psychiatry, which historically has been heavily rooted in the medical model, clustering symptoms and signs into specific disease entities.

ness. These techniques, according to a hospital trade journal, are "the biggest thing to happen to psychiatry since Freud."[42]

PATIENT PROFILING AND THE USE OF TESTS FOR ECONOMIC PLANNING

As diagnostic tests are becoming more inclusive and precise, physicians can project a more complete picture of the diseases that are likely to develop during the course of a person's life. Such "patient profiles" are a way of planning long-term health care. They also provide information with profound implications for a patient's access to medical resources.

Patient profiles are based on both family history and physical and laboratory examinations. Profiling provides a basis for institutions to predict demands for costly medical resources. In the case of third-party payers, the profiles developed through diagnostic tests can be used to exclude certain individuals from coverage or to increase their insurance rates.

Insurers argue that a major criterion for defining baseline insurability is that diseases be beyond the control and expectation of the insured.[43] When risks are known, premiums must be adjusted accordingly. When new tests such as the ELISA (Enzyme-Linked Immunosorbent Assay) test for AIDS identify high-risk groups, insurers create new classifications to decide whether a person is insurable and how to set the rates. If predictive genetic tests are available, insurers expect to use them. "If testing is prohibited," says Bob Hunter, president of the National Insurance Consumer Organization, "people in standard health will subsidize those at risk."[44]

The debates over AIDS testing suggest the pressure for predictive patient profiles as a basis for underwriting. Insurers have insisted on testing for thc AIDS antibody, arguing that it is necessary to avoid the economic collapse of the industry. According to a 1987 survey by the OTA, 86 percent of responding commercial insurers have either screened or plan to screen for HIV infection.[45] Some simply incorporate questions in health history applications that might reveal high risk for AIDS; others "redline" those whose zip codes indicate they live in high-risk areas of a city. But most insurance companies require a physician's statement or routinely test applicants themselves.

Analyzing the use of AIDS tests by insurers, a lawyer observes the well-established right of insurance companies to require evidence of insurability. "The most fundamental principle of insurance—charging the same premium rate for individuals whose expected risk of loss is the same—requires that insurers and insureds have equal access to any knowledge that has a significant bearing on the assignment of an individual to the appropriate class of risk."[46] The main challenge, he claims, is to identify people in the early stages of a high-risk disease who want to buy insurance. Insurers anticipate that AIDS testing is only the beginning. "Looming on the horizon is the prospect of extensive genetic testing, which may reveal that certain individuals are prone to this or that serious disease."[47]

In a similar vein, the president of the Association of Life Insurance Medical Directors discusses the use of genetic tests for Huntington's disease: "Any information available to an insurance medical director would enter into a decision about either health or life insurance."[48] In some cases such predictive information could ease an insurance burden. For a person known to be at high risk for Huntington's, a negative test would open the possibility of preferred insurance rates.

The technical complexity of administering tests is likely to preclude their direct use in the insurance underwriting process in the near future. However, insurers expect that test results will one day become part of the medical record of applicants, who will have to acknowledge their existence when filling out medical history questionnaires. Moreover, insurers will have access to such information through required statements by physicians.

Insurers organize information about prospective customers through the Medical Information Bureau (MIB). This organization of 700 insurance companies who share information about policy holders serves as a kind of detective agency, protecting its members against concealment of underwriting information. The MIB data bank stores the profiles of about 11 million people. A person who applies for life, health, or disability insurance with one of the member groups signs a form giving the company permission to provide medical information to MIB. Any information available from genetic tests would be included in these files and would be accessible to all member firms.

Predictive tests for most genetic diseases have been available only since the mid-1980s. The probe to identify carriers of the dominant Huntington's gene was not made available to interested medical geneticists until four years after the marker was discovered because its developer, James Gusella, feared the misuse of the information, its emotional effect on the afflicted individual, and its implications for discrimination against those with the condition.[49] The "gene" linked to manic depression has been studied only in a very limited population, the Pennsylvania Amish. Complex but common behavioral disorders such as alcoholism are still identified mainly through observation, interviews, or family history. As diagnostic techniques develop, however, their use will follow directly from the existing strategies employed by health care institutions to enhance efficiency, support decisions, and control costs.

The Growth of Medical Control

New technologies promise significant benefits for consumers, contributing substantially to the quality of health care. The explanations these technologies provide are in great demand by both physicians and patients. Though the results of sophisticated tests are often inconclusive,[50] they can identify problems, reduce uncertainty, and allow people to take preventive measures. The extent to which the information gained from diagnostic technologies actually affects therapeutic decisions remains in question, but both patients and physicians value sophisticated diagnostic techniques for their ability to provide "objective" information. Ironically, the very objectivity of diagnostic tests can undermine the ability of the patient to control medical decisions while they reinforce institutional control.

The use of tests to predict illness and behavior can help health care institutions plan long-term policies, allocate resources, and anticipate needs. Tests also help to minimize the risks and repercussions of malpractice litigation, to meet the constraints of reimbursement policies, and to develop statistical justifications for controversial decisions.

Services that are apparently beneficial to patients have often enhanced institutional control. Recall the complex consequences of the discussions on preventive health in the 1960s. For decades, strategies such as health screening and promotion of lifestyle changes were widely accepted with little controversy, viewed primarily as a means for individuals to control their own health. By the 1980s, however, definitions of healthy lifestyles were used by insurers to control costs and by employers to screen job applicants.[51] Similarly, diagnostic technologies that predict those genetically at risk for specific types of cancer are of value to consumers, who might change their behavior or diet to minimize risk. But health care providers and insurers have a distinct interest in such information as well.

New diagnostic and screening technologies reinforce existing institutional strategies by providing physicians with a sense of certainty and an ability to do something even if a cure is not available. They serve as an objective basis for professional decisions in situations where the interest of patients conflicts with institutional needs. They provide hospitals with tools to control access to beds and facilities and to anticipate demands. They also provide a vehicle for negotiating the outcome of consultations.

Diagnosis defines the boundary of the normal. But the boundaries derived from diagnosis are often malleable, subject to socially constructed interpretations. For those found to be at risk, diagnostic categories may themselves have a social meaning shaped by the needs of social institutions. Medical conceptions of behavior and disease pervade the workplace, the schools, and the courts as these diverse institutions embrace the power of diagnostic prediction. They are placing a new emphasis on objective and predictive information about the individuals within their domain, and they are interpreting such information to meet their immediate social and economic needs.

CHAPTER 5

Testing in the Workplace: Predicting Performance and Health

IN 1980 Anthony Mazzochi, then director of the Oil, Chemical and Atomic Workers (OCAW), predicted that the 1980s would be "a decade of genetic struggles in the workplace."[1] He was referring to the controversial practices of several companies that were using genetic tests to screen workers for their potential susceptibility to disease from exposure to toxic chemicals.

Identifying susceptibility to chemical exposure, or any potential health risk reflected in an individual's genetic makeup, is an important application of the advancing knowledge of genetics. Faced with absenteeism and the provision of costly medical services, as well as regulatory pressures and litigation, companies involved in manufacturing chemical products have considerable incentive to predict who might be susceptible to occupational disease. And, as they often serve as insurers and health care producers, employers also have a stake in detecting which workers or job applicants may be predisposed to genetic disease.

Testing is endemic in the workplace. Employers screen workers to select or exclude job applicants, to maintain productivity, and to improve health. Several large chemical companies began to use genetic tests in the 1970s to identify and then exclude employees suspected of being predisposed to illness when exposed to certain chemicals. Such exclusion was a paternalistic measure, designed in the first instance to protect workers' health. At the same time it was far less expensive than making structural changes in the workplace.

There are several ways to minimize the risks of working with toxic chemicals, including changing the organization of work or the physical structure of the workplace. But most protective measures have centered on the individual worker—for example, providing masks or special clothing, developing training programs to change worker behavior, rotating those in high-risk jobs, or excluding people in special categories such as women of childbearing age. As an extension of this emphasis on the individual, identifying and excluding "hypersusceptible" workers is a way to decrease the incidence of occupational illness and the associated costs of compensation.

In their role as insurers, employers have extended their concerns—and their control—beyond the workplace to the personal lifestyles and health profiles of employees and their families. Genetic information can allow cost-effective planning strategies; it is a way to predict and control absenteeism, to avoid compensation claims, and to reduce future medical costs. Today, for the most part, information on worker health is still gathered through medical histories. Genetic screening is not in widespread practice, discouraged in part by the adverse publicity that surrounded the screening of workers in the early 1980s. But from the employer's perspective, testing workers is rational policy. Employers look to improvements in the accuracy and reliability of tests as a way to justify exclusionary practices that might otherwise appear arbitrary or discriminatory and therefore subject to political and legal challenge.

This chapter first lays out the economic and regulatory pressures that encourage diagnostic screening in the workplace. We then look at current strategies to select and maintain a healthy and productive work force, suggesting the tendencies to seek explanations for worker health problems in the biological or psychological condition of the individual and to look for solutions through exclusion. In this context, we discuss how emerging biological techniques can be used in ways that extend the control of employers beyond the workplace and into the personal lives of workers.

Pressures on the Firm

Corporations these days play multiple roles in their relationship to their employees. Beyond their capacity as employers, they also act as insurers and health care providers through their benefit packages. Employer concerns, then, begin with occupational hazards and their effect on productivity. But they extend beyond the workplace to the general health of employees and the future cost of their medical care.

The pressures to reduce occupational illness are most obvious in the many industries using commercial chemicals to create products such as basic chemicals, pharmaceuticals, plastics, paints, and pesticides. Workers potentially exposed to chemicals include the 4.6 million employees of the chemical and chemical products industries, along with millions of others who work with chemicals in a broad range of occupations and firms.[2]

Perceptions vary about the extent of the risk involved in working with toxic chemicals, but most people believe that employers should be responsible for diseases caused by on-the-job exposure to dangerous substances.[3] This view is reflected in government regulation. Federal laws such as the Occupational Safety and Health Act and the Toxic Substances Control Act require industry to assume responsibility for protecting employees. The Occupational Safety and Health Administration's (OSHA) regulations, enacted "to assure so far as possible every working man and woman in the nation safe and healthful working conditions," require the Secretary of Labor to set standards assuring that employees are protected from threats to their health "to the extent feasible economically and technologically."[4] Screening individuals and excluding those who are possibly susceptible protects workers while minimizing the costs of meeting regulations.

The litigious nature of the regulatory environment encourages employee testing as industries seek ways to protect their economic interests. Administrative decision making in the United States is an adversarial process, delegating ultimate responsibility to the courts. The litigation model for resolving regulatory dilemmas emphasizes "finding the facts" and "establishing the truth," a process often ill-suited to dealing with the problem of scientific uncertainty so characteristic of occupational health.[5] In this adversarial context, industries look with favor on diagnostic techniques that seem to provide "hard" scientific data.

Genetic screening has also been encouraged by an obscure federal regulation, Title XXIX of the Code of Federal Regulations (1974), which contains an ambiguous statement (in part 1910) that all workers are required to take a pre-assignment physical exam which is to include "a personal history of the employee and family and occupational background, including genetic and environmental factors."[6] The rule was adopted apparently without consulting government agency officials and, according to an OSHA directive, "does not require the exclusion of otherwise qualified employees from jobs on the basis of genetic testing." Despite concern among OSHA officials, the 1985 Code of Federal Regulations still includes the reference to genetic assessment, justifying consideration of genetic factors in determining who can or cannot work.

Competitive pressures for efficiency and productivity reinforce the appeal of scientific testing. Industries view tests for drugs, for AIDS, or for genetic information as a means of providing help for workers while reducing medical costs, decreasing absenteeism, and avoiding the costs of changing the structure of the workplace. Shifting the locus of responsibility for occupational health from the workplace to the worker reduces the need for costly cleanup programs.

The threat of litigation has further encouraged corporate health-screening policies. Traditional tort recovery is precluded under most worker compensation schemes, but workers still have several possible grounds for litigation.[7*] These include tort action against third parties, such as a chemical manufacturer. In 1981, for example, the Johns Manville Corporation faced 16,500 lawsuits from workers harmed by asbestos. Workers can also sue on behalf of a fetus harmed because of exposure to chemicals; and they can sue corporate physicians.

The potential liability of corporate medical services encourages diagnostic screening. As industries assume the role of health care provider, they also assume a duty to warn. The courts have considered in-house medical programs as a business endeavor intended to fulfill corporate purposes (that is, saving money). Thus companies are liable for the professional acts of the physicians they employ.[8] In particular, corporate physicians can be held

*Compensation laws were developed to substitute predictable administration programs for the more arbitrary decisions of the courts. They do not allow workers to sue their employers directly for compensation, but instead provide social insurance against workplace accidents without regard to fault. However, compensation laws were designed for problems of safety and do not allow proper assessment of the more ambiguous damages associated with occupational disease. This has encouraged efforts to get around their limitations through litigation.

liable for injuries resulting from inadequate diagnosis and failure to warn employees about potential harm. They may, for example, be liable for the actions of impaired employees such as bus drivers or airline pilots who endanger others. If a company has not used an available test to detect a potential problem, it may be considered at fault. In this legal context, the duty to warn encourages the use of diagnostic tests that can identify an individual's susceptibility to chemical exposure.

THE COMPANY AS INSURER

Corporate insurance plans reinforce the appeal of acquiring predictive information that would allow them to reduce costs. About 97 percent of medium and large firms offer health insurance as a benefit. Payment for employee benefits exceeded 39 percent of total payroll costs in 1986, and over 21.1 percent of that total went for medical benefits.[9] Many large firms insure themselves or sponsor their own HMOs. In states where companies purchase insurance from private carriers, premiums are experience-rated. Higher claims mean higher costs. Corporate health costs have been rising at the rate of 20 to 30 percent a year. Among the fastest rising costs in the health care inflation are mental health expenses, growing at an estimated rate of 25 percent a year. One industry consultant believes that "corporate America has got to rethink its whole attitude toward mental illness." Another calls mental illness "one of the hottest areas for cost containment."[10]

The case of Honeywell Corporation illustrates the corporate response to escalating health costs. Its benefit office pays claims for its 65,000 American employees through six insurance carriers and pays premiums to 125 health maintenance organizations and preferred-provider organizations. Between 1978 and 1983 payment for health services rose 19 percent a year, prompting the company to appoint a task force to investigate the problem. The result was a plan to shift the emphasis from "managing health care costs to managing health." Honeywell is, according to a company spokesman, focusing on "identifying risks, personal and environmental, that will affect health negatively in the present and in the future." This includes collecting data on the personal and medical history of its employees.[11]

In their role as employers, companies face legal, regulatory, and economic pressures that encourage the use of diagnostic techniques to identify and then exclude those workers predisposed to illness from exposure to chemicals. As succinctly stated by a *Forbes* journalist: "If people vary in their genetic susceptibility to the disease, then some will have a high probability of getting ill from exposure to chemicals in the workplace. Screen out those workers and you cut occupational illness."[12] In the role of insurer, a company's rational response to rising health costs is to use genetic tests to predict those predisposed to conditions that might impose future burdens on the company or compromise future productivity of the worker. The focus of concern has, therefore, shifted to include the personal habits, the general health status, and even the genetic makeup of employees.

Strategies

In January 1978 American Cyanamid announced a policy to remove from its Willow Island, West Virginia, plant all fertile women working with certain toxic substances. Specifically, the policy barred female employees of childbearing age (defined as between the ages of sixteen and fifty)[13] from contact with any chemicals that the company medical director believed could cause birth defects in fetuses, the only exception being women who could prove they had been surgically sterilized.[14] The medical personnel explained to employees that the procedure of "buttonhole surgery" was simple and could be obtained locally in several places. The company's medical insurance would pay for the sterilization procedure, and sick leave would be provided to those who chose to undergo the surgery. Once the fetal-protection policy was fully implemented, those who failed to undergo surgical sterilization could no longer work except in a few low-paying janitorial jobs. Five women were sterilized. Those who did not undergo sterilization were transferred out of their former jobs, and their pay and benefits were reduced to correspond to their new positions.

Scientists know little about the effect of most chemicals on the embryo/fetus during the first few months of pregnancy. Nor do they understand the reproductive effects of exposure on men, but data indicate that many chemicals affect male as well as female reproductive functions and cause birth defects among the children of male workers. Nonetheless, corporate officials at American Cyanamid first considered applying their policy only to women who were already pregnant, and then rejected this idea because they felt that damage to the embryo could occur before women were aware of their pregnancy. They also decided not to consider setting low permissible exposure limits, but simply to ban all fertile women. Moreover, American Cyanamid implemented the policy without addressing the possibility of alternatives such as engineering controls, substitution of chemicals, the use of personal protective equipment, or job rotation.

In January 1980 thirteen women and a labor union representing the workers filed suit under Title VII of the Civil Rights Act of 1964, charging American Cyanamid with job discrimination. It was settled out of court.[15] However, a second issue had been raised a year earlier by OSHA, citing American Cyanamid for violating the general duty clause of the OSHA Act of 1970, which required that companies furnish employees a place of employment free from recognized hazards. An administrative law judge assigned to the case granted the company's motion for summary judgment. The OSHA's Health Safety Review Commission affirmed the judge's decision claiming that the hazards requiring sterilization for fetal protection were not cognizable under the OSHA Act.[16] The Oil, Chemical and Atomic Workers Union, which was a party to the earlier proceedings as an intervenor, appealed the Commission's decision to the D.C. Circuit Court of Appeals but failed to overturn it.[17]

The issues raised in the American Cyanamid case illustrate four strategies that are characteristic of the corporate response to occupational health problems.

1) The tendency, even in the absence of definitive scientific evidence, to place responsibility for problems on the individual worker rather than on the conditions of the workplace.
2) The establishment of in-house medical services responsible for defining the health status of workers.
3) The effort to predict who may be prone to future illness.
4) The development of policies to exclude workers or to increase control over their private lives.

These strategies meet the institutional needs of companies pressed by litigation and regulation as well as economic and social forces. They set the background for corporate receptivity to tests that identify the genetic susceptibility of individual workers.

PLACING RESPONSIBILITY ON THE WORKER

In newspaper "advertorials," the Chemical Manufacturers Association and major chemical companies regularly advertise steps taken by the chemical industry to create a healthier working environment. These measures range from protective masks and clothing to better ventilation, from protective devices to engineering controls. Labor and management have engaged in a long-term debate over the relative value of these approaches to worker protection. Minimizing risk through structural alterations of the workplace or changes in operating conditions or practices can be very costly. Thus economic rationality has encouraged industry to place responsibility for health on the individual worker rather than on the firm, and therefore to favor personal protective devices over engineering controls.[18]

This perspective assumes that workers, rather than defective equipment or management practices, are at fault in most accidents, that health hazards have been substantially controlled, and that management adequately tests chemical substances for potential risk before exposing workers. If workers are inadvertently exposed to toxic substances, it is assumed to be a result of carelessness, failure to follow correct procedures, or unwillingness to use precautions. If workers get sick, it is often attributed to personal habits such as smoking or an unhealthy lifestyle. More recently, it has also been attributed to genetic predisposition to disease.[19]

Such views have long dominated the literature on occupational health and safety, and are often accepted as facts rather than as representations of an industrial perspective. An article on occupational psychiatry in the 1966 *American Handbook of Psychiatry* illustrates how industry's view can be translated into professional assumptions. It associated the pulmonary insufficiency ("pneumoconiosis," "emphysema," "chronic bronchitis") of workers to dynamic diagnoses such as depressive reaction, anxiety reaction, and psychophysiological reaction.[20]

Another view began to emerge in public discourse in the late 1960s as labor activists attributed occupational illness to problems within the control of management. Assuming a basic right to health, they attributed chemical hazards to poor design of equipment or operational procedures, and argued that work should be made safe for all employees. In the absence of definitive scientific evidence about the relative causes of occupational illness, however, questions of responsibility are usually left to the discretion of the company doctor.

EMPOWERING CORPORATE MEDICAL SERVICES

Corporate physicians play a strategic role in establishing responsibility for problems of health and safety in the workplace and in defining the health status of individual workers. Corporate medical programs were first set up by companies in forestry, mining, and railroads to provide rudimentary medical services in remote areas. Between 1900 and 1920 they were introduced into industries to screen workers in order to ensure higher labor efficiency, to reduce absenteeism, to monitor compensation claims, and to establish fitness to work. American industry employs about four thousand occupational physicians and twenty thousand industrial nurses, who work mainly in large firms. Most companies with more than fifty thousand employees have full-time M.D.s and medical departments. DuPont, for example, employs seventy full-time and fifty part-time physicians, as well as two thousand registered nurses in its plants throughout the country.[21] Among the physicians are an increasing number of occupational psychologists, hired to assess the suitability of job candidates on the basis of psychological tests.

Company doctors act as mediators between employers and individual employees. But most of them are themselves employees, hired by a firm as a cost-control strategy and dependent on the firm for their wages and the conditions of their employment. Accountable to corporate management and in some cases formally part of management, often positioned as vice-presidents for health, they are responsible mainly for using medical criteria to select and maintain a productive work force. Testifying before a congressional committee on the responsibilities of medical departments, a spokesman from Southern Railroad succinctly described their obligations: company physicians are "to provide the industry with applicants who are as near perfect physical specimens as it is possible for us to find."[22]

The work of corporate physicians includes both medical and administrative tasks. Their medical work includes pre-employment medical examination or screening of job applicants, routine examination of workers to monitor changes in their health, provision of health care services and referral to specialized doctors, and evaluation of illnesses and complaints. Their administrative tasks include collecting and interpreting data on the general level of health and safety in the plant and maintaining records on individual workers and the work force. They act as advisers to management, reporting on matters of health and safety, and they are often called upon to testify in worker compensation cases, usually representing the company's interests. In effect, they practice a form of medical adjudication, evaluating medical fitness for promotions, eligibility for insurance, responsibility in tort liability cases, and justification for absences from work.

These tasks are relatively straightforward in cases of accidents or obvious physical disabilities. The job of corporate physicians, however, has been complicated by the occupational health problems associated with the use of chemicals in the workplace. Their judgments take place in a context of technical uncertainty that limits precise interpretation of symptoms and precludes the definitive association of illness with work. Many of the available diagnostic techniques are primitive, leaving open broad possibilities for subjective judgment. The inherent uncertainties of evaluation are compounded by ambiguities in the corporate physician's role. Caught between two clients, the individual worker and the corporation, these physicians are a kind of double agent. They are employed as part of a strategy to meet the institutional needs of their employers, a strategy that focuses much of the responsibility for health and safety on the worker. But as physicians, their code of ethics clearly mandates their professional obligation to their individual patients. And these patients may have claims against the firm.

Corporate physicians resolve the ambiguities inherent in their position by asserting professional and scientific objectivity.[23] Their professional ideology emphasizes the neutrality and objectivity of medical expertise. Refined diagnostic tests that provide a scientific basis for their judgments are welcome as a means to clarify the ambiguities of their role.

PREDICTING PRONENESS TO DISEASE

American firms have been engaged in what has been called "a frenzy of inspecting, detecting, selecting and rejecting."[24] Testing is extending well beyond the general pre-employment medical examinations intended to evaluate fitness for work. Tests are given for drugs, for AIDS, for personal habits and personality characteristics, for honesty and—most recently—for seeking out those workers who may be prone to future illness.[25] Workers who smoke, who are obese, who have high blood pressure, who use drugs, or, as in the American Cyanamid case, who are women of childbearing age, have been the target of policies based not on present job performance but on susceptibility to future health-related problems.[26] These workers may find they are ineligible for company insurance plans; they may be required to attend remedial programs such as antismoking or stress-reduction programs, or they may be denied a job. Hundreds of corporate programs are offered to help employees deal with alcoholism, drug abuse, smoking, or stress. The approach is not to reduce organizational sources of stress, but to train individuals to better cope through changes in their personal lifestyle.

Predicting a worker's proneness to future health problems is part of the medical examination frequently required of new job applicants. Medical examinations in the workplace have a long tradition. The worksite is an efficient place to reach large numbers of people and to find latent or hidden health conditions that would not otherwise be detected.[27] Thus public health officials used worksites to screen for tuberculosis in the early part of the century. By 1917, 10 percent of the three hundred largest corporations in the United States were giving routine medical examinations to their employees and to their applicants to select those workers most fit for particular jobs.[28]

In 1985, 49 percent of U.S. employers required pre-employment medical exams, an increase from 38.5 percent in the early 1970s.* Various predictive tests have been included in these exams, ranging from psychological tests for potential executives to lower back X-rays for construction workers. Even dubious testing procedures persist. About a million pre-employment back X-rays are taken on asymptomatic individuals despite evidence that only 2 percent of those screened out ever develop serious back trouble.[30] A manual of testing for employers lists hundreds of available instruments to test the psychological states, values, attitudes, judgment, stress, alienation, needs, trustworthiness, honesty, stability, and other personality characteristics of prospective employees. Some of these tests predict potential levels of accomplishment or productivity; others seek to identify maladjusted individuals, defined as "alcoholic, inadequate and immature" (one such test is called the Manson Evaluation) or to predict "those most likely to be responsible for theft."[31] Until they were restricted by Labor Department regulations in December 1988, 2 million lie detector tests were given to job applicants and employees every year despite questions about their validity and legislative efforts to limit their use.

The appeal of predictive testing has been evident in the increased screening for drug use in American corporations. Most drug tests are urine analyses, or in some cases hair analyses, that

*Preplacement medical exams were required for 19.2 percent of employees in small firms (18–250 workers), 48.9 percent of employees in medium-size firms (250–500 workers), and 83.3 percent of employees in firms with over 500 workers.[29]

simply measure drug use. Brain wave tests are being developed to measure impairment from drug use, assuming it will predict work performance. Screening programs experience a large number of false positives (caused, for example, by cold medications, diet aids, and, in the case of one paroled prisoner, poppy-seed bagels [see page 46]).

There is considerable skepticism about the value of such tests as an effective deterrent to drug use, and there are legal concerns about the invasion of privacy and intrusion on other constitutional rights. Nevertheless, drug testing enjoys widespread credibility and currency, and is not limited to those with visible job performance problems. Companies use drug tests both to select new employees and to anticipate and control potential problems among existing employees. In 1987 *Chemical and Engineering News* surveyed twenty-two major chemical producers to find that they all had instituted routine drug testing procedures for new job applicants.[32] Those testing positive in the course of a regular pre-employment physical exam were rejected. Many companies were also subjecting existing employees to random or periodic testing. Faced with the possibility of legal challenge, however, the companies offered rehabilitation treatment to those who tested positive for the first time.

It is predicted that drug testing of new job applicants as a basis for exclusion will increase despite objections from civil libertarians and evidence of limited accuracy. In February 1987 the Reagan administration announced a sweeping drug detection program requiring about a million federal workers to submit to monitored urine tests. Estimates are that over half of the Fortune 500 companies are presently testing their employees. In March 1987 a market research firm estimated that more than 5 million employees and job applicants would be routinely tested each year. Revenues from the sale of drug diagnostic tests reached $100 million in 1986 and are predicted to be $250 million in 1988. Drug testing is, according to diagnostic firms, a "growth market."[33]

Driven by social as well as economic pressures, employers have also urged widespread testing for AIDS. Advocates of AIDS testing want to exclude those who test HIV-positive, arguing that it is necessary to protect co-workers and customers as well as the infected worker in order to promote productivity and to control health care costs. In a 1987 poll, 42 percent of respondents said all employees should be tested for AIDS.[34]

The trend toward increased testing in the workplace has been very controversial. Much of the criticism focuses on the inaccuracies of the tests and their limits as predictors of behavior. Most drug tests, for example, measure the use of drugs but not the effect of drug use on job performance. X-rays and tests intended to predict proneness to disease identify people who may never develop the anticipated conditions. Nevertheless, predictive screening is increasingly viewed as an appropriate and defensible means to evaluate job applicants. The conventional wisdom, stated in the *Journal of Occupational Medicine*, is that, "In terms of business economy, this method fosters sound practices since it can reduce the rate of injuries or illness . . . lead to reduced absenteeism, increased productivity, and decreased expenditures for workers' compensation and group health insurance."[35]

EXCLUDING PEOPLE FROM WORK

In 1980 the Equal Employment Opportunity Commission (EEOC) filed forty complaints from people who had been barred from jobs because of alleged hypersusceptibility to chemicals. They were mostly women, excluded under fetal-protection policies designed to prevent exposure of an embryo or fetus to toxic substances in the workplace. Exclusionary policies, such as the one behind the American Cyanamid incident, have been adopted at many corporations, especially in the petrochemical sector.[36] In a 1980 survey of chemical companies, conducted by the trade journal *Chemical and Engineering News*, every company maintained that, in principle, no women biologically capable of having children should be exposed to substances posing a direct risk to the health and viability of the unborn child.[37]

If consistently applied, this policy would have an extraordinary impact on the employment of women. Women's groups and civil rights organizations have responded with hostility to fetal-protection policies, viewing them as an excuse to reverse civil rights gains by reinforcing the stereotypes that had long blocked women from full participation in the labor force. They believe the designation of "susceptibility" will be interpreted in terms of existing social biases, citing the lack of evidence to support fetal-protection policies. And they note the limited attention given to the reproductive hazards of men in similar circumstances, despite the widely recognized contribution of males to negative reproductive outcomes. They object to defining all women in terms of their biological role and thereby limiting their employment opportunities.[38]

In response to complaints, the EEOC proposed interpretive guidelines on reproductive hazards and discriminatory employer policies. Concluding that the scientific evidence supporting fetal-protection policies was limited, it sought to restrict the circumstances under which women could be denied employment to include pregnant women only. It did not, however, provide any protection for women transferred or removed from their jobs. Women's groups attacked the guidelines for being too limited. Industrial spokesmen, fearing legal liability, also opposed the guidelines. Referring to the fetus as "an uninvited visitor" with claims above those of the mother, they argued that a woman's obligation was to provide "room and board," a "safe and healthy prenatal environment."[39] They defended exclusionary policies as beneficial to the fetus and necessary to protect industry against tort liability; the EEOC guidelines, they claimed, denied businesses the right to protect themselves. A representative of American Cyanamid stated the case in revealing language: "The ideal is that the workplace has to be safe for everyone. . . . In the real world that's totally unachievable without emasculating the chemical industry."[40] Not surprisingly, in January 1981 the Reagan administration withdrew the proposed EEOC guidelines.

Subsequently the OTA reviewed the literature on reproduction and the workplace.[41] Its report emphasized that there was no biological basis for assuming that either a fetus or a woman is any more susceptible to harm from most chemicals than a male. OSHA has found it necessary to regulate only three substances because of their potential hazards to reproductive health. In these cases it did not recommend the exclusion of workers, but rather the use of engineering controls, protective devices, or temporary removal of pregnant workers. Moreover, neither OSHA nor the National Institute for Occupational Safety and Health (NIOSH) has found any reason to place pregnant women in a higher risk category than other workers. In October 1988 the EEOC ruled that employers may not dismiss or refuse to hire women because of suspicion that exposure to chemicals could cause reproductive damage, unless there is objective scientific evidence that the hazard affects the offspring through the female parent.

Fetal-protection policies have received a lot of publicity because of the dramatic nature of forced sterilization, the obvious challenge to civil rights legislation, and the response of an organized women's movement. But these policies are only one example of a broader tendency to exclude workers who are currently able to perform their jobs but who run a risk of future illness. There are some legal barriers to such exclusionary practice. In *Sterling Transit Co. v. Fair Employment Practice Commission* (1977) an employer fired a truck driver who had a congenital, but not disabling, back condition. The court ruled that the "mere possibility that the employee might endanger his health some time in the future was inadequate justification for the employer's actions."[42] This implies that the injury must be very likely to occur before exclusion can be justified.

Similarly, in *Office of Federal Contract Compliance Programs v. E. E. Black Ltd.* (1980) the court held that under the terms of the Federal Rehabilitation Act of 1973 only an individual's *present* capability to do the job could be considered in a hiring decision, and that potential ability (or inability) to perform the requirements of the job at some future date was not a valid consideration.[43] The court gave a hypothetical example of the conditions under which a possible future illness would be a legitimate basis for exclusion: if an otherwise qualified person had a 90 percent chance of having a heart attack if placed in a certain job, then it would be consistent with the "business necessity exception" to the Rehabilitation Act for an employer to refuse to hire that person.

Aside from avoiding the stigma attached to labeling asymptomatic people as handicapped, these rulings offer only limited protection. The court did not, for example, provide a standard for determining how likely a future illness must be in order to justify legal exclusion under the Rehabilitation Act's business necessity exception. Thus, by implication, these cases leave open the possibility of legal exclusion in the case of an imminent disease or risk, if a disease or disability can be predicted with reasonable certainty and the exclusion justified as a business necessity. In this context, refinements of diagnostic procedures appeal to corporations as a means to minimize the threat of litigation over exclusionary practices.

Current institutional strategies, then, include assigning responsibility for health to individual workers, trying to predict those who are prone to future illness, and developing exclusionary practices and policies. This is the context in which new biological technologies of genetic diagnosis are being introduced into the workplace.

New Biological Testing Techniques

The field of "ecogenetics," based on the idea that genetic variation should be considered in assessing responses to environmental agents, emerged in the 1960s and gained credibility throughout the 1970s.[44] A National Academy of Sciences Report in 1975 devoted a special section to the role of genetic metabolic errors as predisposing factors in the development of toxicity from occupational or environmental pollutants.[45] The report listed ninety-two human genetic disorders thought to predispose affected people to toxic effects. In response, a number of industries began to use such knowledge as a screening tool in their effort to deal with problems of occupational health. They have employed two basic types of genetic tests: chromosomal monitoring and genetic screening.

CHROMOSOMAL MONITORING

Chromosomal monitoring is employed to assess whether exposure to workplace chemicals or radiation has caused chromosomal damage. Monitoring programs involve evaluating blood and urine samples from newly hired workers to look for chromosome breakage and to establish a baseline for the individual. Existing chromosome damage could be due to parental exposure to mutagens, previous exposure of the individual in an occupational or environmental setting, or natural mutation. Once a baseline for the individual is established, later samples are periodically collected and compared. Presumably any change in the amount or type of chromosomal aberration could indicate damage resulting from workplace exposure to toxic substances. Any consistent changes revealed by the monitoring of a working group could alert management to an exposure problem and a need for engineering controls in the working environment, greater use of personal protection devices, or the substitution of chemicals in order to reduce exposure levels.

Monitoring currently relies on cytogenetic techniques that look for chromosomal aberrations in blood or urine samples as an indication of damage to the DNA. This is an indirect method, based on the assumption that abnormalities in blood cells reflect DNA mutations. Noncytogenetic techniques are being developed that directly assess damage to the DNA itself. These will eventually become more accurate diagnostic tools than the present, uncertain methods. While the value of chromosomal monitoring has sound theoretical and experimental grounding, epidemiological studies using such monitoring techniques are inconclusive. Some studies have established a dose-response relationship between increased exposure to a chemical agent and increased chromosomal damage, but none definitively show a correlation between chromosomal damage and greater risk of disease. As long as the evidence of their value remains inconclusive, employers are reluctant to undertake costly monitoring programs. They have argued that monitoring may cause unnecessary alarm among workers and lead to stricter regulations and requirements for expensive changes in equipment. Labor interests, by contrast, favor monitoring as a means of detecting exposure hazards for all workers.[46]

GENETIC SCREENING

Employers have shown greater interest in the use of genetic screening techniques than in chromosomal monitoring.[47] Genetic screening is a one-time procedure designed to detect workers with specific genetic traits or variant alleles that might predispose them to illness when exposed to particular chemical agents. The OTA has characterized genetic diagnostics as "a science truly in its infancy" and concluded that tests "do not meet established scientific criteria for routine use in an occupational setting."[48] The current state of knowledge does not explain differential susceptibility, which may be influenced by development and aging processes, nutritional status, and other variables besides genetic traits. According to the OTA, the data on the correlation between given genetic traits and risk for disease are simply not extensive enough to draw predictive conclusions. Nevertheless, the data are suggestive and employers are using genetic screening as part of pre-employment medical exams.

Employers favor screening as a cost-effective protection for individual workers: an inexpensive way of fulfilling their duty to inform workers about potential health risks. At the same time, screening may minimize the risks of future liability, trim the costs of insurance and compensation claims, and avoid structural changes in the workplace. An OTA survey of the Fortune 500 companies in 1982 found that 17 of the 366 companies that responded had used genetic screening and 59 intended to do so.[49] Those corporations using the tests said that they had selected employees for screening on the basis of job category or ethnicity. They had taken a variety of actions on the basis of this screening: informing the employees of a potential health problem, transferring them, or denying employment to potentially susceptible individuals.

Labor groups have responded to genetic screening practices with outrage. From their perspective, genetic screening is a discriminatory measure that allows industry to select employees to fit the workplace rather than making the workplace fit for employees. Negative media coverage of the American Cyanamid case and a 1980 *New York Times* series on genetic screening led to political activity among both labor and women's groups and then to congressional hearings.[50] Labor activists in particular called the identification of "hypersusceptible" individuals a form of workplace discrimination.

Following this publicity, corporations became cautious about announcing their policies. Only five major companies continued to admit performing any kind of genetic testing. However, corporate interest in diagnostic technologies that would divert compensation claims and financially debilitating lawsuits remained high, especially after the lawsuits filed by asbestos workers against the Johns Manville Corporation. According to a 1986 investigative report in the *New York Times*, many major corporations were again considering implementing screening programs.[51]*

The appeal of genetic screening techniques is consistent with the corporate discourse on occupational health. The chemical industry, for example, embraces the goal of health and safety. But believing that there are necessary constraints on creating a "zero-risk" environment, it emphasizes that decisions about worker protection must be made in an economic context. In effect, health and safety are economic commodities. Accordingly, the language of efficiency enters the discourse on health. How can society best invest its limited resources in this area? How can health and safety be protected in cost-effective ways? How can we balance demands for increased safety against the cost of meeting them? From the perspective of an individual corporation, a rational economic approach is to seek predictive measures that would exclude high-risk individuals and therefore avoid changes in the workplace for the benefit of relatively few employees. Genetic screening fits into this context. Monitoring, with its implications for changing the organization or physical environment of work, is a less attractive alternative.

The principle that a susceptible or vulnerable class of workers can be treated differently because there are "scientific reasons" for discriminatory exclusion has taken on political meaning because of the association of genetic susceptibility with ethnicity. Susceptibility to toxic chemicals is linked to the genetic conditions associated with particular ethnic groups. Predisposing genetic characteristics may leave individuals vulnerable to environ-

*A new OTA survey of genetic screening practices is under way, and its results will be available in 1990.

mental disease, including sickle-cell anemia, thalassemia, and a dozen other heritable traits. Susceptibility is most commonly associated with sickle-cell anemia, a mono-genetic disorder caused by the substitution of a single amino acid at one locus on the hemoglobin chain. A person with the anemia produces an abnormal hemoglobin molecule (HbS instead of HbA). Anemia is found in 1 out of every 400 to 600 American blacks, or 0.2 percent of the American black population. Sickle-cell anemia is highly debilitating, and it is therefore unlikely that an individual would reach employment age and remain unaware of having the disease. But 1 of every 10 to 12 American blacks is a carrier of the trait. Usually asymptomatic, a carrier will experience sickness only when blood oxygen is greatly reduced. Under most circumstances the health hazards of carrying the trait are minimal or nonexistent, and individuals have no way of knowing whether they are carriers.

A person with black ancestors might want to be tested for the presence of the sickle-cell trait in order to make family planning decisions. But employers have screened job applicants for the trait in order to exclude those at risk from jobs requiring contact with certain chemical compounds.[52] In fact, this practice lacks strong scientific justification. While most screening in the workplace has been for the sickle-cell trait, few people in the United States have had occupational health problems associated with this condition. According to the OTA survey, "the attention given to sickle-cell anemia has led to a significant number of incidents in which individual blacks have borne the cost of actions based on mistaken judgments by employers and others."[53] Nevertheless, if genetic tests are to be reasonably accurate, they must focus on groups that have predisposing characteristics. For, as discussed in chapter 3, the predictive value of such tests is very low when they are used to screen the general population. Indeed, if all employees in a firm were to be screened, the test would be virtually useless because of the large number of false positives that would occur.

If accuracy requires screening a high-risk population, namely, those already suspected of having the trait, corporations are faced with a dilemma. In order to gain accurate test results in a cost-effective testing program, they would need to determine who was likely to be susceptible before administering the test. In the case of the sickle-cell trait, this is possible; those most likely to have the condition are ethnically identifiable. But the ethical problem of singling out blacks for testing and perhaps exclusion is obvious, and critics have associated genetic screening with racial discrimination.[54*]

Such concerns are not new. In 1969, four army recruits with the sickle-cell trait died while training at relatively high altitude. The Department of the Navy then promulgated a ruling that all recruits must be tested for hemoglobin S. Though studies of naval recruits indicated that those with the sickle-cell trait were at no special risk, the navy's policy disqualified them from a number of occupations. Ten years later, the Air Force Academy expelled several black men with the trait on grounds that their health might be endangered by the rigorous training program. The exclusion was strongly criticized as unfairly restricting opportunities for blacks. Subsequent studies have differed considerably about the actual risks. Some have found a causal relationship between the trait and the few incidents of death from exertion; others have denied it. A 1987 study suggests that blacks with the trait are at somewhat higher risk. However, the policy question behind the twenty years of controversy was whether suspected problems should be met by excluding those with the trait or by adapting military training and conditions of service to their needs.[55]

The same question emerged in the 1980s in response to the growing use of genetic screening in the workplace. The discussion at congressional hearings illustrates the way a policy that had first been implemented for the purpose of protecting black work-

*Four states—Florida, Louisiana, New Jersey, and North Carolina—have laws banning genetic testing by employers. These grew out of the debates over sickle-cell tests.

ers became viewed as a form of discrimination and a method of social control.

DuPont began its testing program at the request of black employees, with the intention of giving them the results for use in family planning decisions. Under the provisions of the program, the test results were retained in the employee's medical file on a confidential basis and were not intended for use as the basis for employment decisions. A DuPont physician specializing in occupational medicine testified before a congressional subcommittee that the test for sickle-cell trait at DuPont was "purely for the education and edification of the individual involved."[56]

This may have been the original intent, but upon further questioning at the hearing a spokesman from DuPont's public relations department admitted that the company did not have a policy for explaining the test or its results to employees. Thus it appeared that the screening program had been transformed from a potential information source to help employees make personal decisions into a tool to help the company implement biased employment policies.

Further investigation of corporations using genetic screening techniques found that few medical directors had systematic data on the results of genetic tests or their effectiveness. They had no clear idea of how the results were used in the workplace or of how many workers had been denied employment or given alternative job placements because of genetic susceptibility.[57] Nor were there coherent policies for maintaining confidentiality.

Interpreting occupational health in terms of heritable traits of the individual worker eases the economic and regulatory pressures that corporations face due to the prevalence of toxic substances in the workplace. As predictive tests become more refined, for example, through the development of noncytogenetic screening techniques, they are likely to be more widely used, even though the relation to actual illness or job performance may not be demonstrable. As diagnostic techniques develop for the prediction of late-onset diseases, they too will be tempting additions to such programs. Some testing has become a part of employee "wellness" programs. A firm called Focus Technologies in Washington, D.C., is testing employee groups for evidence of a range of common diseases and genetic disorders. The program involves computer analysis of family histories and thirty-two different blood tests.[58] The purpose is to provide health education for employees that is tailored to their individual needs.

Industry justifies testing and the exclusion of high-risk workers in terms of protecting their health. The executive director of the New York Business Group on Health, a corporation of three hundred businesses concerned about health and productivity, provides the rationale: "If an employee is an epileptic and subject to fainting spells, does it make any sense to put him to work on a scaffold?"[59]

The business press clearly expresses the temptation to use genetic tests. In 1987 *Industry Week* observed that genetic testing, despite uncertainties, can indicate which diseases a person is likely to get. "Such knowledge could save business costs and increase employee efficiency." The growing use of testing for drugs and for AIDS has laid the foundation for acceptance of such tests. According to Bruce Karrh, the corporate medical director of DuPont, drug testing "has probably broken down the psychological barrier [for genetic tests]."[60]

Genetic Struggles

The "genetic struggles" in the workplace of the 1980s are likely to accelerate in the 1990s. Most controversy to date has centered on the implications for racial and sexual discrimination. But as the ability to determine genetic criteria becomes more refined and the use of predictive tests increases, other controversial questions will emerge.

The purpose of genetic screening is prediction. Screening is not an assessment of the present state of health, but a probability statement about future risk. Like diagnostic screening in other contexts (for example, the schools), such tests can serve as the basis for judgments about a person's future ability to perform a job. This has several implications. People with chronic health problems, such as epilepsy, now have problems getting work. Could not genetic diagnostic techniques create a growing class of unemployables, not on the basis of existing symptoms but on the anticipation of possible future symptoms? In a competitive economic environment, industry must try to select the "best" employees on the basis of both potential productivity and future health. If future medical risk becomes a criterion, individuals could find themselves on genetic blacklists, classified as unfit for work. While medical records are private, the protection of confidentiality is in some circumstances limited.* Thus, some union leaders are concerned that workers will bear a "genetic scarlet letter," that they will become "lepers" or "genetic untouchables."[62]

Tests are not infallible. Moreover, there may be little correlation between positive tests and impaired performance. Yet data from tests are compelling: though a person may have no symptoms, a diagnosed predisposition to a disease can itself be perceived as a kind of abnormality, a disability, a disease. A person can be transformed into a patient without actual manifestations of disease. Indeed, in the discussions of genetic screening, "hypersusceptibility" is treated as a handicap justifying exclusion, even though proneness to disease has nothing to do with present performance.

Exclusion from work has a pejorative social meaning. Work-

*A 1986 case suggests that while disclosure of medical information about mental competence of employees (in this case a bus driver) would be an invasion of privacy, compelling public need can outweigh issues of confidentiality.[61]

related evaluations carry over to other aspects of life, so that a person deemed unfit to work is often regarded as unfit or inadequate as a person. The fact that screening categories are based on specific scientific criteria and refer only to potential future disabilities does not reduce the broad negative implications of such designations.

Employers and company physicians now ask most job applicants whether they have any physical condition that might impair their ability to work. They now test them for psychological attributes that might predispose them to behavior viewed as unsuitable on the job. Why not inquire about genetic predispositions? As the sophistication and scope of testing for genetic markers increase, corporations could also justify routine testing of prospective employees for their predisposition to late-onset conditions that might someday impair them. Would corporate resources such as special training programs be invested in those individuals with "bad" markers? Can the right to be employed depend on having the "right" genes? Companies insure not only workers but their dependents as well. As genetic linkage tests become available to diagnose predisposition to disease, can companies use such tests to deny insuring employees or their families?

Routine negotiation and conflict resolution methods offer relatively few ways to deal with such questions. Collective bargaining is an unlikely forum. Unions tend to focus on bread-and-butter issues of wages, hours, and benefits. Health and safety issues have traditionally been less important. Moreover, collective bargaining agreements do not address decisions about the criteria for employment.

Legislative protection available to those subject to exclusion on the basis of genetic testing is also limited. The Civil Rights Act of 1964 prohibits job discrimination, but the Supreme Court has held that apparently discriminatory practices are permissible if they are based on "business necessity." Similarly, the Rehabilitation Act of 1973 in theory would protect against discrimination based on genetic screening. However, it too is limited if policies can be shown to be business necessities. The fact that few laws directly address the problem of genetic screening allows corporations considerable independence in interpreting diagnostic information.

Actions against those employees already working is limited by laws, such as the Employment Retirement Income Security Act (ERISA), requiring compensation and health benefits for sick and injured workers. But these protections—applying to employees, not applicants—tend to encourage more rigorous pre-employment screening and the exclusion of those likely to be predisposed to illness.

Industry has tended to deny the adversarial implications of occupational health policies. A rhetoric of consensus pervades the corporate discourse. Health is discussed as a common interest shared by management and labor alike: "All of us are in favor of good health," says Elizabeth Whelan, an epidemiologist and critic of federal regulatory policy.[63] Diagnostic screens are viewed as a humane measure to protect potentially vulnerable workers against unnecessary risk. And, with increased accuracy and reliability, genetic tests are seen as neutral, scientific tools providing the means to protect worker health. As one scientist predicts, "We are now at the threshold . . . of a really new era, a block of genetic information that can be used to create a new chapter in genetic medicine."[64]

More precise information about individual vulnerability would help define both ethically acceptable and economically rational policies; for there are certainly cases in which exclusion would be a reasonable protective measure. However, even with improved techniques, decisions about genetic susceptibility cannot be assumed to be neutral. They can be linked to existing stratification systems, that is, to race, gender, and social class.[65] And they are implemented within an existing structure of power in the workplace. Corporate managers and the medical professionals accountable to them control diagnostic procedures, interpret the data from screening tests for their predictive value, and decide on the appropriate course of action. And they are necessarily guided by the need for economy and productivity.

Despite increasing precision, predicting the level of "risky" susceptibility, defining "problematic" exposure levels, and resolving the many uncertain dimensions of diagnostic interpretation still involve many discretionary policy decisions. What are the thresholds of dangerous exposure? How much certainty is necessary before excluding a person from a job? It is left to the company doctor to translate ambiguous data into unambiguous corporate policy. In this discretionary framework, testing, when used as a criterion for social action, can become a strategy to extend corporate power and control beyond the workplace into the personal life of workers; beyond their manifest fitness into their genetic profile; beyond their present productivity into their future prospects for a productive and healthy life.

CHAPTER 6

The Schools: Testing for Learning Problems

SINCE the mid-nineteenth century educational reformers have debated the relative influence of heredity and environment on the success of students in schools. Some attributed the learning problems of students to school systems that were trapped in the "crust of sterility and rigidity" or estranged from the "life of the working-class community." Others attributed problems to "lack of vital force" or "inherited tendencies to vice."[1]

Such debates have persisted. In the social climate of the 1960s, explanations of academic failure focused on environmental sources of behavioral and educational problems. While some education analysts were developing measures of intelligence that assumed inherent biological capabilities, the prevailing view stressed the importance of socioeconomic factors and family deprivation in shaping educational performance. This emphasis on environmental causation was striking in the school psychology/learning disability literature of the 1960s. For example, learning difficulties were frequently attributed to "personality maladjustment"[2] or the school environment.[3] Child psychiatrist Irving

Harris attributed all learning problems to home life: "In the home, most of the reasons for failure in learning and success in learning will concern the mother," who might cause learning problems if she were ambitious, jealous, tyrannical, divorced, or emotionally unstable, but most of all if she were to "enjoy working outside the home."[4]

In the 1980s such social explanations have been virtually replaced by the tendency to attribute both positive and problematic qualities of individual students to their own biology. It is increasingly assumed that both learning and behavior problems reflect biological deficits, lying less in a student's environment or social situation than in his or her brain. Even social or environmental stresses are frequently explained in terms of their biological effects: a deprived environment depresses neuronal growth, or a child is devastated by family trauma because he is "genetically susceptible" to environmental stress.

Reinforcing such beliefs are hundreds of popular books and magazines on child care. An article in *Child* magazine called "Cracking the Code" advises parents to track the language progress of their babies as an indicator of brain development. It includes a description of a test called CLAMS (Clinical Linguistic and Auditory Milestone Scale) to monitor infants from birth to two years. "Get as much information as you can about their brain development," advises a psychologist quoted in the article; for learning problems start in infancy, reflecting the organization of the brain.[5]

From our perspective, the interesting issues raised by this recourse to biological explanations lie not in their accuracy or on whether behavioral or learning problems in fact have identifiable biological correlates. Clearly some learning or behavioral problems are the result of genetic, neurological, or biochemical conditions. And, just as clearly, uncertainties about the developmental stages of childhood have cast doubts on biological reductionism. Our interest is rather in the implications of the growing faith in biological explanations for the way tests are increasingly used to categorize children in schools.

Responsible for assessing and classifying children, for channeling them toward future social roles, schools are inevitably sensitive to cultural biases, public expectations, and political demands. More than any other institution, schools depend on testing. They use tests to classify children in ways that reflect both prevailing social forces and current scientific theories about the nature of intelligence and the causes of behavioral problems.

The system of classification and channeling—the way students are managed in a school—reflects the economic and ideological context in which school systems operate. Changes in this context over the past few decades have profoundly affected administrative approaches to learning and behavioral problems, contributing to enthusiasm for biological reductionism. Biological approaches, reinforced by the availability of diagnostic instruments, meet institutional needs at a time when greater accountability is demanded of the schools.

Biological explanations have played a central role in educational theory for most of the twentieth century. But in the 1980s diagnostic techniques have refined the capacity to detect and anticipate ever more minute differences in individual ability. They have also provided scientific justification for the fact that the number of students labeled as learning-disabled tripled between 1976 and 1982, reflecting the economic and political incentives of federal legislation designed to accommodate handicapped children.[6] Increasingly sensitive diagnoses have implications for both students and educational institutions. Just as diagnoses in the clinical context define therapeutic approaches, so too they shape educational solutions to learning problems in the schools. And they reinforce the power of educational institutions to define problems, to classify their students, and therefore to shape their future roles.

The structure of screening, diagnosis, individual testing, and channeling in the school system of the 1980s rests on the expectation that students will benefit. Yet this structure also serves institutional needs. Quantitative test data, for example, can satisfy external pressures for accountability. Indeed, the social and economic pressures facing school systems are encouraging more emphasis on testing, prediction, and biological explanation, with implications for the power of school professionals and the rights of students and their parents.

Social Pressures on the Schools

American schools face growing pressure to be accountable for their educational practices, and they are responding with a renewed emphasis on standardized tests. The demands on school systems reflect increased public awareness of widespread illiteracy and general disappointment with incompetent educators. But these demands also reflect the sense that American industry is falling behind in a global race for efficiency and profit, for schools are viewed as responsible for creating a competitive labor force. Indeed, the demands placed on educators in the 1980s mirror the larger social tensions in American society.

ACCOUNTABILITY

Americans have often turned to the school system to address social problems, viewing it as a force that could resolve conflicts external to its core responsibility.[7] The "melting pot" theme of American education in the Progressive Era was a response to social tensions over immigration. During a period of growing nativism, when immigrants were conspicuously not assimilating, schools were called upon to transform young foreigners into middle-class, all-American kids. Similarly, the 1954 *Brown v.*

Board of Education decision mandating desegregation of public schools came before the full thrust of the civil rights movement, at a time when blacks and whites in the South were not yet sharing water fountains, public libraries, or neighborhoods. In the struggle to change a well-entrenched social system, the school was sent to the front line.

The public schools in the 1980s are called upon to deal with different social needs. It is the school system rather than special hospitals or sanitariums that must meet the needs of mentally retarded and handicapped children. In addition, schools are expected to respond to the declining leadership of American industry in the international marketplace. The rhetoric of science literacy emphasizes the dire economic consequences of "technophobia" in American schools and blames the schools for the widening gap between the labor force and the needs of a technological society.[8]

Public expectations about the social role of education were strikingly apparent in the 1983 report of President Reagan's National Commission on Excellence in Education, *A Nation at Risk*. The Commission concluded that the state of the nation's educational system was a direct threat to national security and economic growth, and faulted schools for failing to prepare workers for a technological society, for having "squandered the gains made in the wake of the Sputnik challenge," and for mediocrity in teaching and administration.[9]

These conflicting pressures leave the American educational system with time-consuming, inflexible, and often inconsistent methods of accountability. Administrators must meet the needs of the physically and mentally handicapped while preparing the most gifted to save America from international economic decline. They must mainstream as many students as possible, but provide special educational services to at least 12 percent of the student population. Such demands have prompted educators to seek unambiguous "scientific" or technological measures of student abilities and performance. These measures help schools classify and channel students appropriately, and facilitate efficient planning. They also allow schools to explain educational failure, thereby deflecting blame and reducing responsibility.

MEETING LEGISLATIVE DEMANDS

The response of school districts to the passage of the Aid to All Handicapped Children Act of 1975, PL94–142, illustrates how external pressures can promote increased testing. Intended to guarantee free public education to all handicapped or learning-disabled children, this Act provides for funds to support special learning programs and mandates accounting procedures to justify the funding. It formalized the procedures for placing students in special education, stipulating that the federal government will provide a specified amount—approximately $2,500—for each learning-disabled or special education student, for up to 12 percent of a district's students. No additional funds are available beyond the 12 percent. The legislation requires parental involvement in the classification of their children, who must be placed in the least restrictive environment possible.

Schools have responded to the legislation with the intention of providing opportunities for children in need of extra help. The law offers the possibility of much-needed money for personnel and equipment. Its funding provisions encourage districts to seek out learning-disabled or handicapped children. Because the amount of money increases with the severity of the handicap, administrators in schools with federal funds are motivated to maintain a full complement of handicapped students in each category in order to maximize support. A loss of the federal funding could affect equipment, classroom aides, and salaries. And so, with the best of intentions, administrators try to meet the provisions of the law. This inevitably influences the way students are classified. As one study suggests, "Students are more likely to be placed in a special education program that is short of students than in a program that is full. We found that students are not likely to be removed from a special education program and returned to the mainstream until there are others to replace them."[10]

Local and state pressure on school districts can be equally demanding. Many states mandate course requirements, bilingual education, specific textbooks, or classroom standards. These demands converge with internal organizational needs for classroom order, efficiency, staff communication, and economy, creating pressure to reduce diagnostic ambiguity. Standardized tests, with their aura of scientific objectivity, have long played a role in helping school districts respond to these complex institutional demands.

Current Strategies

STANDARDIZED TESTS

Testing students has long been an important way to gauge performance and meet demands for accountability and compliance in American schools. Tests assess the efficiency and performance of the school system as well as the abilities of individual students. Despite a long history of criticism concerning the validity of tests and their potential for social misuse, testing has persisted as the central assessment technique of the education system. A substantial literature exploring the use of tests suggests how intelligence testing in particular has served extrascientific purposes—reinforcing racial beliefs, justifying social inequalities, and promoting individual careers.[11] When Alfred Binet developed tests to measure the intelligence and educational attainment of students just after the turn of the century, he called attention to the limited information provided by test scores. He viewed IQ test scores only as a pragmatic means to identify children in need of help, and he believed they were neither indicative of innate intelligence nor tied to any theory of the intellect.[12] However, in that period pervaded by hereditarian ideas, mental tests were soon accepted as revealing genetic intelligence and embraced as a "scientific" tool of classification.

Intelligence tests were widely used by army personnel in World War I, and within a few years after the war, the Alpha and Beta tests developed by Harvard psychologist R. M. Yerkes were being administered to millions of American schoolchildren each year.[13] Testing solved well-recognized educational problems of student placement and classification, and encouraged a growing emphasis on individual differences among students. The enthusiasm for testing in the schools in the 1920s created a new profession, the school psychologist, then called the "tester." It also defined the field of psychometrics as the science of measuring mental traits, abilities, and processes.

In 1935, New York was the first state to certify school psychologists, and in the same year the University of Illinois offered the first training program. School psychologists were trained not in psychology, however, but in test administration. It was not until the 1950s that the profession sought to extend beyond this limited role. The professional literature began to urge practitioners to take on clinical responsibilities and educational activities, and to influence administrative policies. Nonetheless, they have continued to serve mainly as "testers" in the schools, responsible for measuring intelligence.

The reification of "intelligence" as a single, innate, and fixed characteristic that could be scientifically measured continued largely unquestioned through the 1950s. Standardized tests provided educators with numerical indicators of ability, and so justified the general bias in favor of middle-class white children, who tended to score higher on IQ tests than immigrant, poor, and black children. In the late 1960s and early 1970s, however, school psychologists became aware of problems of racial bias and labeling intrinsic to IQ test performance. Earlier, differences in the scores of black children had been interpreted as evidence of fundamental racial inferiority. By the 1960s educators recognized the effect of cultural and social background on test scores. IQ tests came under suspicion, fueling a bitter controversy over the relationship of genetics to IQ.[14] According to one professional handbook, by the late 1960s most school psychologists had abandoned the belief that IQ tests measured innate, fixed, inherited intelligence.[15]

In response, the California Supreme Court in 1982 prohibited state educators from administering IQ tests to blacks. This decision suggested the inadequacies of testing to measure intelligence, but the notion of a mental quality called "intelligence" that is inherited in measurable ways has persisted in both educational theory and practice. Indeed, controversial theories associating genetics and IQ still guide some research. Arthur Jensen, a veteran of the 1960s debates over IQ testing, has continued his experiments throughout the 1980s, trying to quantify inbreeding as a factor in intelligence.[16] Also reflecting such beliefs, a 1985 study analyzing the standardized test scores of twins reports that educational achievement is 70 percent genetic and 30 percent environmental. The authors conclude that genetic factors determine general ability while environmental factors "channel it into proficiency."[17]

Suspicion of IQ tests has turned attention to the legitimacy of the instrument rather than the legitimacy of what it measures. Some psychometricians began to improve the tests, searching for culture-free instruments.* By the 1970s standardized tests began to reflect the emerging scientific metaphors of the brain. No

*For example, they tried to eliminate culture-loaded questions and to develop new techniques for "norming" exams, such as using sample groups that represent the population to be tested to construct the instruments.

longer seen as a single entity, intelligence came to be defined in terms of neurological circuitry and was assumed to include many separate functions. An assignment of IQ is no longer considered sufficient to classify a child. Thus tests today include psychological indicators, and school records contain extensive psychological information about individuals and their families. This approach, however, still focuses on individuals, minimizing the importance of environmental factors. And it still rests on defining statistical norms against which behavior and intelligence can be measured and assessed.

In their widely read book, *The Myth of the Hyperactive Child*, Peter Schrag and Diane Divoky described the range of tests that children were exposed to in the 1970s: "They cover not only the familiar categories (learning disabled, minimal brain damaged, emotionally handicapped, visually handicapped) but also include scales rating children for impulse control, intellectuality, withdrawal and social behavior. . . . The incentive is to find problems." They go on to suggest that "it is normal to be watched, tested and treated . . . and the associated techniques—screens, drugs, behavior modification, special programs—all serve the purpose of legitimizing and enlarging the power of institutions over individuals."[18]

In practice, school systems today use three general types of tests to evaluate and classify students. First, mass-administered tests provide aggregate scores on educational attainment. Focusing on "normal" children, these are important for the comparison and assessment of different schools and school systems. Districts want to improve the overall average scores attained in these tests, since such scores reflect on the quality of the education provided by the district. The district's average score can be improved, presumably, by improving instruction, but also restricting the pool of students tested. If poor performers are not counted in the mainstream classroom, their scores will not pull down the school or district average.

Poor performers can be kept out of the mainstream testing population through a second level of screening. Many American schools administer initial screenings as a quick way to assess the hundreds of children as they enter the school system. These limited screening tests, administered in five to fifteen minutes, provide very little data on which to assess a student, and problems can easily be missed. But they are an efficient way of catching some problem students early. Those performing poorly on the screening tests are then subjected to the third kind of testing, the individualized battery of cognitive tests administered by a school psychologist. These tests are expected to pinpoint specific neurological or perceptual problems and to provide guidance for placement or treatment of the student.

Recent advances in testing presume to locate both cognitive capabilities and learning disabilities ever more precisely in the biological configurations of students' brains. Thus, educators today classify students under elaborate taxonomies which are presumed to provide information about the underlying biological cause of their learning or behavior problems. In 1981 the National Joint Committee on Learning Disability, formed by six professional organizations concerned with learning disabilities, described such problems as "intrinsic to the individual" and "presumed" to be organic in origin.[19] Refined testing techniques reflect this presumption; educators assume that evidence about neurobiological functioning will contribute to the understanding of dysfunctional learning.

Explaining individual differences in terms of the inherent biological state of the student can be an effective strategy as school systems struggle to meet external and internal pressures. Such explanations, reinforced by "objective" tests, can limit the school's responsibility for students' failures. They ease parent-school relations by diminishing parental and teacher blame for poor student performance. They bolster professional authority by framing student problems in "scientific" and hence unassailable terms that many parents feel incapable of questioning. And they justify medical solutions, particularly drugs, for student problems which could be addressed through more traditional methods such as removal to different special educational courses or psychological and social counseling services.

PRESCRIBING DRUGS

In the fall of 1985 the elementary school in Derry, New Hampshire, referred first-grader Casey Jessop to a pediatric neurologist because he was overactive and disruptive in the classroom.[20] On the basis of a brief examination (ten minutes, according to Casey's parents) the doctor diagnosed hyperkinesis and prescribed Ritalin (methylphenidate), a psychostimulant frequently prescribed for hyperactive children as a strategy to enhance both the student's capacity to learn and the school's ability to maintain order.

Ritalin has met with considerable criticism for having dubious benefits and harmful medical side effects. Casey took the prescribed drug for eighteen months. He became, according to his parents, much quieter in school, but at home he showed severe side effects, including sleeplessness and weight loss. His "antsy" behavior, they felt, was leading to much more severe problems. The school, however, kept him in a mainstream class, offering at that time no special programs. While taking the drug Casey's classroom behavior conformed to teacher expectations. Thus school officials insisted that his parents comply with the recommendations of the neurologist, who reconfirmed the diagnosis and the prescription of Ritalin, adding medication to help Casey sleep.

By April 9, 1987, distressed at Casey's continued problems ("He looked like an addict") and seeing no change in his ability to learn, his parents took him off Ritalin. The school, however, insisted he continue medication and referred the child to the neurological center at Boston Childrens' Hospital for an objective evaluation. This center gave Casey a series of tests and confirmed the diagnosis of hyperkinesis, but called it attention deficit disorder. They prescribed an individual educational and counseling program which included the use of Ritalin or a substitute drug. Casey's parents refused to give him the drug and the school psychologist refused to provide him special services without the drug. Indeed, the school board suspended Casey from the classroom and brought his parents to a hearing before an officer of the State Department of Education. School officials testified at the hearing that without Ritalin, Casey was defiant, disruptive, and verbally abusive, but that he was manageable with the medication. The parents agreed the child had problems, but insisted on counseling without medication. The State Education officer upheld the school's position, ruling that Casey must take medication to remain in public school. The parents appealed to a federal court, and the ACLU is litigating the case. Meanwhile Casey is in a private school for emotionally disturbed children, off of Ritalin, and doing well behaviorally and academically.

Similar incidents have occurred since the early 1970s, when Ritalin first came into widespread use. Many parents have complained that by treating students' problems as biological deficits, schools fail to explore alternative explanations: the family dynamics, the classroom context, the effect of interaction between teachers and peers. Satisfied if a child's behavior while on medication conforms to the demands of the classroom, teachers may fail to consider the total child and ignore the side effects of medication. When assuming a biological model of behavioral problems, they tend to minimize the importance of social or behavioral interventions. And when a medication is apparently effective in helping a child conform to expected classroom behavior, it serves to reinforce the initial diagnosis, validating assumptions about biological causation.

Ritalin's effect on disruptive children was first reported in 1959. The stimulant, the researchers observed, produces a paradoxical calmness and compliance in highly active and agitated children. In the 1960s and 1970s the drug was widely prescribed for active boys (90 percent of hyperactive children are male) who caused classroom problems.

Ritalin can dramatically change the classroom behavior of overactive children like Casey, allowing them to control their behavior sufficiently to learn. Many studies of Ritalin's educational effects, however, have concluded that the drug improves neither attention span nor comprehension, that it is used indiscriminately by some educators mainly for its sedating effect on troublesome young boys with too much energy, and that it can serve as a control device for the benefit of teachers.[21] This was the recurring criticism of Ritalin throughout the 1970s, when a series of lawsuits led to a temporary decline in the use of the drug.

A redefinition of "hyperactivity" in the 1980s deflated the criticism by reframing the problem. The word had been used to describe a behavior problem within the social context of the classroom, but in 1980 the third edition of the *Diagnostic and Statistical Manual of Mental Disorders* (*DSM III*), published by the American Psychiatric Association, replaced it with a new term: "attention deficit disorder." This term describes an internal brain state of the student, independent of the social group. The revised *DSM IIIR*, published in 1987, further differentiates among behavioral disorders, adding several categories such as "stereotypy/habit disorders" and "expressive writing disorders." This refined classification implies that such problems are developmental disturbances—mental disorders rather than simply a result of inattention. As a respected guide to the diagnosis of mental disorders, the *DSM* has had substantial influence. Attention deficit disorder has become "the most common behavior problem of school-age children," believed to affect an estimated 3 to 10 percent of the 45 million children in school.[22]

By the mid-1980s the use of Ritalin also greatly increased. In 1987 the U.S. Drug Enforcement Administration, which regulates the production of controlled substances, doubled its production ceiling to 2,682 kilograms, reflecting the increase in Ritalin sales.[23] The increase has both reflected and stimulated diagnoses that focus on the neurological basis of attention deficits.

Research in neuropsychology suggests that attention deficit disorders involve the slow processing of information, a diagnosis that makes some sense of Ritalin's "paradoxical" effect. Children with the disorder are characterized as suffering from a slower than normal information-processing system, which prevents them from understanding or making sense of classroom instruction. This leads them to be bored and restless. Psychostimulants such as Ritalin are claimed to "shorten central nervous system processing time and reduce variability; thus messages can be understood better and individuals can stay on-task longer with medication."* This characterization of the disorder recognizes hyperactive behavior as "a relatively minor secondary problem."[24] In fact, a diagnosis of attention deficit disorder no longer requires the manifestation of hyperactive classroom behavior. This new formulation of the problem has justified the use of the drug for treating a larger pool of students.

In the case of hyperactivity, a problem in classroom dynamics (an overactive student disrupting the class) has been reframed as a problem located specifically within the brain of the individual student (an "attention deficit disorder"). The *DSM III* reformulation of hyperactivity as an internal, biological aberration of the child justified the shift in explanation. The effectiveness of Ritalin in easing behavioral problems in the classroom encouraged the attribution of a broader range of learning problems to brain biochemistry, enlarging the number of students who might conceivably benefit from using the drug. This, in turn, shifted attention farther from the social context of the classroom and toward the biology of the individual student.

*Note that a number of studies suggest normal children show similar reactions.

In fact, every child and every school situation is unique. No matter what the general trend, some children will be helped by a treatment intervention and some will not; each case must ideally be judged on its merits. Yet it is difficult for most schools to provide such individual attention. Thus, as student behaviors are increasingly classified in biological terms, the use of pharmacological solutions will rise. Frustration with the difficulties of student assessment has converged with substantial public pressure to make such medical solutions attractive as a strategy to meet institutional needs. At the same time, new diagnostic technologies promise much greater precision in identifying the causes of learning and behavior problems and offer the "objective" data that might help resolve conflicts such as those underlying Casey Jessop's case.

Emerging Biological Explanations and Diagnostic Advances

In the late 1960s and early 1970s, expectations about the power of psychological testing to predict the behavior of children evoked some remarkable proposals. In December 1969 Arnold Hutschneider, a physician, proposed to President Nixon that the government should engage in mass testing of all six-to-eight-year-old children in order to detect those with violent and homicidal tendencies. Nixon's domestic affairs adviser questioned the wisdom of identifying "pre-delinquents" to control future criminal behavior, and it was never implemented.[25] But the psychological assumptions of the 1970s easily slid into biological explanation of learning and behavior problems, encouraging efforts to identify those who might be predisposed to learning or behavior problems prior to their manifestations. By the 1980s tests were promising to detect "soft signs" of neurological damage and to identify future problems stemming from genetic abnormalities.[26]

Research in neuropsychology, behavioral genetics, and medical genetics has had a growing influence on educational theory and practice. Based on the assumption that all neurological dysfunctions not explainable by trauma are innate, part of the genetic endowment of the individual, these fields seek correlations between deviant behavior and abnormal brain activity. Psychologists are extending biological explanations to the personality traits of children. Developmental psychologist Jerome Kagan has observed a correspondence between shyness and inherited variation in the physiological characteristics of the brain. He speculates that "children we call inhibited belong to a qualitatively distinct category of infants who were born with a lower threshold for limbic-hypothalamic arousal to unexpected changes in the environment."[27] He believes it is important to identify such children through physiological measures. In an interview in *Harvard Magazine*, Kagan suggested that "other traits are similarly determined through biology. The sense of morality, for instance."[28] Other researchers are focusing on inborn personality traits that predispose children to be risk takers, suggesting that the tendency to abuse alcohol and drugs might be predictable. Researchers have suggested that children predisposed to be risk takers should be identified at an early age so that their energy can be rechanneled.[29] While many people question such assumptions, they are nonetheless shaping the progression of new diagnostic technologies and their application in the schools.

Professional discussions about reading difficulties also illustrate the shift to biological explanations. Until the 1960s most reading difficulties were viewed not as illness but as the bottom end of the normal curve of reading ability. Educators tested vision and hearing to assess slow readers but did not consider them to be abnormal, and the condition was not usually attributed to biological deficits. Today, however, learning disabilities fall within the professional realm of neuropsychologists, who use several diagnostic approaches to explore children's genetic predisposition to reading problems.[30] Behavioral studies concentrate on families and twins in an effort to find genetic relationships. Cytogenetic studies focus on boys with an extra X chromosome.* These XXY children frequently have difficulty reading, and the research seeks correlations between their disability and their chromosomal disorders.[31] A further set of studies focuses on genetic linkages, selecting families with reading disabilities over several generations in order to establish the link between the occurrence of the disability and a chromosomal structural abnormality. All these studies are hampered by diagnostic imprecision; in effect, the multiple causes of reading disabilities and their complex and diverse manifestations preclude diagnostic clarity. Nevertheless, most studies of learning disabilities today are based on biological assumptions.

This shift to the biological model is in part a result of advances in diagnostic technologies. In 1927 neuropathologist Samuel Orton was one of the first scientists to propose neurological causes for those reading difficulties that were later classified as dyslexia. Yet Orton had no way of specifying the nature of neurological damage. The effort to locate such problems within the brain began seriously only in the 1960s and 1970s with the development of more sophisticated technologies: computerized electroencephalographs (CEEGs) to monitor activity in the brain; the Sensation Cognition Computer to measure cortical activity; the Neural Efficiency Analyzer to gauge the speed with which the brain processes information. In the 1980s the development of new brain imaging technologies has given neurologists and neuropsychologists a central role in diagnosing and recommending treatment of students with learning problems.[32]

*Some children with an extra Y chromosome have been found to have reading problems as well.

The knowledge developed from sophisticated imaging techniques has shaped the way the problems of students are defined and classified. Brain imaging can reveal structural or chemical abnormalities, and genetic testing can explain learning problems in boys with, for example, the Fragile X syndrome.[33] While school administrators and counselors do not themselves use the most advanced technologies, their approach to student behavioral problems is framed by the medical discourse in which such technologies play a prominent role. The experimental development of new diagnostic technologies has reinforced the belief that behavioral problems are rooted in biological deficits that only remain to be detected when technologies become available. Neuroscientists, for example, fully expect to be able to find signs of underlying intelligence and behavior problems by testing cognitive functions. According to the National Institute of Mental Health, new technologies are refining the understanding of thought processes: "It is conceivable that intelligence and behavioral style may become definable and recognizable by characteristic patterns of brain activity."[34]

Faced with the enthusiasm of researchers who seek to promote applications, and feeling the need for efficient solutions, educators tend to accept the promises of science. They are willing to assign biological causes to the behavioral and intellectual problems of children even when biological tests are not yet available. Indeed, research is quickly absorbed into educational practice. One obvious measure of this can be found in education journals, where articles make broad promises about the application of brain research not only to understanding learning difficulties but also to tracing predisposition to drug abuse, violence, and other behaviors that present problems for the schools.[35]

Each new development in brain research has spurred efforts to establish biological explanations and categories of learning problems and to devise fast, accurate diagnostic tests to recognize these conditions in students. For example, theories of brain lateralization revived the use of EEGs in diagnosing the source of learning difficulties. EEGs had been used since the late 1940s to identify abnormalities in learning-disabled children. By the late 1960s, the test was debunked as a predictor of achievement, though still recommended as a diagnostic tool. But EEGs gained renewed credibility with the development of neurometrics, a field devoted to the study of brain dysfunctions through statistical analyses of the electrophysiological features of the brain. In neurometrics, computers are used to analyze huge amounts of EEG data on the differential responses of subjects to various stimuli. The intention, according to E. Roy John, a researcher in this area, is to establish a set of norms against which to measure individuals.[36] Deviations found to exist in learning-disabled children can provide objective signs of brain dysfunction even in asymptomatic children.

Further developments around EEGs involved mapping activities in different parts of the brain. Brain Electrical Activity Mapping (BEAM) organizes EEG information to indicate which parts of the brain are activated during certain tasks. This technique is used to compare the brain activity of dyslexic children with the patterns in normal children.

A number of new technologies extend these diagnostic trends. Computerized brain scans that follow the flow of blood through the brain while people read or think can detect differences in the brains of those with dyslexia. Similarly, by tracing radioactive substances, PET scans can visualize brain activity, allowing investigators to observe the process of learning as individuals perform tasks that use specific regions of the brain. With the help of CEEGs, school psychologists have been able to identify schoolchildren who deviate from the norm, correlating aberrant activity in the frontal region of the brain with their behavioral patterns.[37] Neuroscientists are also trying to understand attention processes by identifying the brain structures that underlie them. They expect to link attention disorders with specific brain dysfunctions, and then to develop tests to identify the dysfunctional structures and to predict those children likely to develop disorders.

Dyslexia has been a major research focus. This disorder is usually diagnosed on the basis of reading difficulties that cannot be explained by obvious problems such as poor eyesight. The mapping of electrical activity in the brain has provided evidence that dyslexia is associated with consistently aberrant function in two well-defined regions of the cerebral cortex. This, it is believed, establishes the disability as inherently organic and opens the way to identifying potentially dyslexic children at a very early age, before they show any behavioral manifestations.[38] Meanwhile, a geneticist has uncovered a marker for dyslexia, and hopes to be able to screen preschoolers in order to provide early remedial training and compensation strategies before the disability is manifested.[39]

Superconducting quantum interference devices (SQUIDs), coming into use as a means to diagnose brain functions and disorders, have particular applications in schools. According to Dr. Samuel Williamson, one of their developers, they permit detailed testing of such functions as sensory perception, short- and long-term memory, and the process of paying attention, all without the use of invasive probes. SQUID could become a part of routine physical examinations: "It's no longer a question of whether such an annual brain check will be developed, it's a question of when."[40]

School psychologists currently diagnose learning disabilities on the basis of cognitive and psychometric tests. But while cognitive tests can suggest functional brain problems, they do not by themselves provide sufficient information to "ground" the diagnosis in any meaningful way in the physical structure or biochemistry of a child's brain. The experimental work in neurometrics and the existence of sophisticated imaging techniques lend credibility to biological explanations. It has shaped the discourse on learning disabilities and framed the interpretation of the findings from "softer" cognitive tests.

Frustration with the limits of existing tests makes more precise brain-imaging technologies attractive for several reasons. Many children do have serious learning disabilities, and accurate diagnosis can facilitate the development of progressive remedial programs. However, receptivity to new technologies also reflects the broad public appeal of simplistic solutions to ambiguous and difficult problems. Witness the encouraging promises headlined by the press: "Brain Studies Shed Light on Disorders" or "Solving the Mysteries of the Mind."[41]

Classification and Control

Schools have long differentiated and classified students through diagnostic and evaluative tests. As the ability to detect subtle differences among children has increased, so too has the interest in differentiation. Educators in general are "much more aware now of individual differences in children," according to a school psychologist.[42] Detailed and sensitive information about individual children—their genetic makeup, predisposition to violence and mental illness, brain structure, and susceptibility to disease—serves well-recognized educational needs. Classifying students also serves the administrative needs to meet demands for public accountability, to reduce blame for learning or behavior problems, and to enhance efficiency and economy in the management of education. Technologies that image the brain, assess genetic capabilities, and reveal biochemical states that "cause" behavior allow educators to predict which children will be slow learners, disruptive, handicapped, or difficult in the classroom.

Such predictions may have advantages for both the educator

and the student. For the educator, faced with various administrative pressures, predictive tests save time and increase school efficiency. For the student, predictive tests call attention to difficulties in vision or hearing, spatial perception, language skills, or emotional development. Assessing these problems early can facilitate remedial measures. Special help can be provided before the child has had a chance to grow frustrated, discouraged, or fixed into habits that may later impede the ability to learn.

Biological explanations are useful for quite pragmatic reasons. They organize student behavior patterns in ways that are useful to school systems under pressure to justify funding, to produce high achievers, and to respond to the concerns of parents. Diagnostic taxonomies appear to simplify the recognition of problematic or potentially problematic students, by providing scientific or rigorous standards that minimize the ambiguity of such diagnoses. Biological taxonomies also help school systems meet both external and internal needs, by providing explanations that diminish school responsibility for student failures and meet legal requirements for federal funding. By defining problem students as ill rather than bad, they may help gain public support for special programs. An appeal for support of blameless victims of their own biology may attract more public sympathy than an appeal for "troublemakers" or "juvenile delinquents." Moral concepts are important in American politics. Attributing problems to biological forces over which students have no control gives educators a model population of the worthy or needy who deserve public funding.*

*David Rosner examined the worthy and unworthy poor in nineteenth-century America, suggesting how concepts of worthiness shift with institutional interests. Physicians in private practice adopted strict definitions of "worthiness" to limit the ability of "paying" patients to use "free" charity health services. Trustees of charity hospitals, however, eager to promote the usefulness of such institutions, developed broad criteria for "worthiness" to increase their patient pool. Similarly, in the 1980s concepts of worthiness help qualify victims for public assistance (food stamps, housing, special education). Thus for educators, medical explanations provide a population of blameless and therefore worthy students who thereby qualify for public support.[43]

Given the cultural tendency to attribute social and behavioral problems to the chemical qualities of the brain, it is not surprising that the educational establishment has uncritically accepted the biological model as an explanation of student behavior. But what are the consequences of labeling a child a victim of "lateral imbalance" or "minimal brain dysfunction" rather than a victim of "nutritional deficiency syndrome"? In both cases, remedial help is indicated. But the biological model poses special problems within the context of the school system. The most obvious problems follow from the dilemmas inherent in any form of educational classification: labeling can lead to stigmatization, and testing poses problems of confidentiality and threats to personal privacy. Biological classifications can exacerbate these problems because they imply that the individual lacks control over his or her behavior and is faced with a permanent and immutable disability. While biological explanations remove moral disapproval of behavior problems, they also reduce expectations.

Labels also suggest preferred treatments such as special diet, special education, or drugs, and the proper arena for problem solving, within the school, the legal system, or the larger society. Biological conceptions of learning and behavior problems may encourage uncritical reliance on pharmacological solutions. Children with behavioral problems have been treated with psychostimulants (Ritalin, Dexedrine), antipsychotic agents (Thorazine, haloperidol), antidepressants (Tofranil, Elavil), anti-anxiety agents (Librium, Valium), anti-manic agents (lithium carbonate), and sedatives (phenobarbital, Benadryl).[44] While these must be prescribed by a physician, the school system has become a major source of information in identifying the problematic behavior among children that appears to call for drugs.

Further dilemmas lie in the interpretation of biological tests. In light of the importance of diagnosis to the subsequent education of the children tested, to their future prospects, and to their own self-image, valid interpretation is crucial. Even where technical information is available, student assessments continue to rest mainly on "soft" interpretations. Schools assess students on the basis of three factors: cognitive tests, psychological evaluation, and teacher referral reports. Cognitive tests provide scores that can be documented and used to explain student placement; they are more "scientific" than either psychological assessment or teacher referrals. Yet a number of studies indicate that these tests are not the most important diagnostic tool in student evaluation. The widely used Gesell Development Test, according to one study, "made no significant contribution to the discrimination of promoted and non-promoted children."[45]

The single most important factor influencing evaluation remains the teacher referral which, even in the face of contradictory test data, guides decision makers at the staffing level.[46] Evidence suggests that many teachers have a low tolerance for students with learning and behavioral difficulties and that their referrals primarily reflect concern about classroom order and control.[47] Biological explanations can lend credibility to such armchair diagnoses, encouraging the differentiation and segregation of problematic students when such practices reflect organizational more than educational goals.

Finally, the focus on biology may direct attention away from social sources of learning problems. Analyses of "minimal brain dysfunction" have suggested that the problem may be strongly influenced by social conditions, and that even a child's growth can be affected by environmental deprivation. By focusing attention on the physical state of the brain, educators may neglect the social and economic conditions leading to damage. By dealing with the biological characteristics of individuals, rather than the social causes of problems, schools set up solutions that insulate them from responsibility while enhancing their immediate control.

In a highly critical review of research on learning-disabled children and its application in the schools, clinical psychologist Gerald Coles claims that current approaches are based on "an unswerving postulation that the source of the individual's problem lies within the individual's neurology; a proclivity for finding biological causes; a disregard for experiential explanations; illogical reasoning about the relationship between behavior and the brain; a misinterpretation of symptoms; and a readiness to apply a medical label to superficially diagnosed and insufficiently understood academic problems." He argues that the development of more refined tests and instruments and more sophisticated methods for understanding the brain have produced "a pedagogical and psychological extravaganza."[48] In 1985, 1.8 million children were categorized as learning-disabled. Yet, except for a relatively small number of children, biological tests have failed to provide explanations that are useful in the understanding or resolution of learning difficulties.

Coles's point is that children have a range of neurological differences that interact with social circumstances to affect their ability to learn. Without denying the reality of certain serious neurological dysfunctions, he insists that most children with learning problems do not have significant brain abnormalities. By assuming that neurological deficits underlie all problems, diagnostic techniques have deflected attention from the importance of social interaction in shaping behavior. They therefore have further mystified our understanding of learning difficulties.

An increasing variety of cognitive and biological tests are available to assess, compare, and channel students. They help teachers, school administrators, psychologists, and counselors establish standards and provide legitimacy for their professional decisions. But the use of these diagnostic techniques may have substantial social force beyond the educational context. The school system has access to most children in the society and is traditionally responsible for assessing, categorizing, and channeling them toward future roles. Biological testing gives school professionals the tools to construct a model or standard of a prototypical student against which to measure individuals. It enables them to show where deviations from the standard can be statistically related to subsequent behavioral or learning difficulties. Diagnostic evaluations in the school can feed into other institutions, defining what, in the context of learning ability and social behavior, is normal or pathological, identifying who is biologically constituted to assume certain types of jobs. Diagnostic technologies can help schools meet their own internal needs, but they also empower them in their role as gatekeepers for the larger society.

CHAPTER 7

The Medical Empowering of Legal Institutions

IN 1983 the U.S. Supreme Court addressed the question of due process in admitting psychiatric predictions. In *Barefoot v. Estelle*, a case concerning the sentencing of a convicted murderer,[1] two psychiatrists had offered testimony about Barefoot's potential for rehabilitation based only on hypothetical illustrations of whether such a criminal would be likely to continue to be dangerous. The American Psychiatric Association challenged the validity and reliability of such evidence. Despite the limits of psychiatric prediction and the critical importance of a case involving the death penalty, the Court decided that such evidence was admissible for a sentencing decision. No better information was available at the time. (Barefoot was unable to afford an opposing psychiatric witness and the State was not obligated to provide one.) Moreover, according to the Court, there was no reason to bar such testimony because even lay persons could arrive at the same conclusion.

This case, often cited in the legal literature, suggests the dilemmas created by the lack of hard evidence to guide decisions about the disposition of defendants in the courts. It also explains the growing appeal of scientific evidence from biological tests. Indeed, by late 1985, the courts in California began admitting evidence from PET scans to assess the mental state, the culpability, and the appropriate disposition of criminal defendants.

The courts often call upon psychiatrists to evaluate the mental state and rehabilitation potential of convicted criminals. Such evaluations have serious consequences for decisions about culpability and sentencing. Yet they are frequently inaccurate, uncertain, and open to capricious interpretation. Lacking objective data, judges and juries tend to rely on intuitively plausible indications of accuracy. In a study of expert witnesses, David Faust and Jay Ziskin point to the influence of such factors as "the expert's stated confidence and description of his judgmental processes and powers, and his background, training, experience, and credentials."[2]

Uneasy with the possible prejudice intrinsic to such capricious and judgmental factors, the courts have looked to science to provide more solid insights into human behavior. Criminal law uses scientific expertise to determine whether an individual should be held responsible for violating society's norms or excused as irrational and unable to control his conduct. Tort or negligence law relies on science and technology to provide evidence for judging culpability. Belief in the power of science to provide hard facts shapes decisions about the proper disposition of those responsible for criminal behavior. And, scientific evidence is increasingly valued as a means to enhance the efficiency and effectiveness of overcrowded courts.*

*A 1986 survey of the effect of forensic science in the adjudication of felony cases found that scientific evidence has made relatively little difference in the determination of guilt or innocence compared to confessions and tangible evidence, but it has had the greatest impact at the time of sentencing. Where scientific evidence is offered, convicted defendants are more likely to go to prison and for longer periods. This study also found an

Although the law remains conservative in its use of new information as the basis of judicial decisions, scientific and technological developments, changing public attitudes, and the need for greater efficiency in the administration of justice have led to the greater use of scientific information in the courtroom. In particular, the legal system is responding to scientific trends in genetics and the neurosciences by shifting its concept of acceptable expertise, placing greater emphasis on biological explanations of behavior and scientific test results.

Legal decisions involving diagnostic technologies touch on important and often controversial concepts about the nature of legal evidence. Psychiatric evidence has frequently conflicted with traditional legal norms, posing problems of proof in the courtroom. Yet technologies that appear to provide "hard" facts are compelling. They are gaining importance in assisting judges to determine the admissibility of evidence in legal disputes, and juries to reach their verdicts. The effect of these trends on the outcome of legal decisions is not always clear. In some decisions the use of biological techniques in the courts is appropriate and helpful, for example, when used to affirm the identity of criminals. But there is little evidence that would allow us to generalize about whether medicalizing deviance helps courts make the right decision about criminal responsibility and the disposition of offenders.

Pressures and Demands

Certain pressures on the legal system have encouraged the use of scientific information in both criminal and civil law. Most important is the inadequacy of the insanity defense, with its reliance on

---

increasingly favorable climate for greater use of scientific evidence among prosecutors and jurists.[3]

questionable psychiatric evidence to exculpate offenders from criminal responsibility for their deviant acts. In addition, the administrative strain of burgeoning caseloads encourages the use of evidence that will facilitate decisions. Increasingly, the courts see various forms of biological testing as a way to enhance the efficiency of the legal system and its power to delineate responsibility in complex tort and negligence disputes.

DEFINING CRIMINAL RESPONSIBILITY: THE INSANITY DEFENSE

Insanity can serve as a complete defense for an individual who violates the law.[4] The function of the insanity defense is to identify those offenders who suffer such extreme mental problems that it would be unfair to call them "bad."[5] Because insanity renders these offenders morally nonculpable, they are excused from responsibility and will escape punishment. In assessing criminal responsibility, insanity is a legal, not a medical, term; being mentally ill is not the same as being legally insane. Mental illness such as schizophrenia or manic depression refers to the condition or state of mind of the afflicted individual, but "insanity" is a functional term, linking emotional and cognitive capabilities to specific behavior. Specifically, tests for insanity focus on individuals' ability to conform their behavior to legal requirements (volition), and to understand or appreciate the nature of their conduct (cognition).

The history of the insanity defense has been marked by frequent efforts to relate the legal understanding of insanity to the current scientific and medical understanding of human behavior. It is a complex history, strained by continued tensions between society's need for order and stability and the individual's right to be protected from arbitrary and unconstitutional restrictions.[6] And it has been wracked by conflicts over the definition of insanity and the criteria used to judge the mental state of criminal offenders.

The legal definition of insanity has varied over the years, reflecting society's changing tolerance and expectations as well as scientific understanding. During the eighteenth century sanity was measured by comparing an individual's mental capacity to that of the average fourteen-year-old; if it met or exceeded it, the individual was judged sane. Courts used a "good/evil" test to exculpate those who were unable to distinguish between the two, and a "wild beast" test to excuse those who could neither understand nor remember their actions.

The insanity test was more precisely formulated in the 1840s by the M'Naghten Rule. This rule grew out of a trial for the attempted murder of Sir Robert Peal, then prime minister of England, in which his private secretary was accidentally murdered in his place. The perpetrator, shown during the trial to possess an extreme paranoiac personality, was acquitted as not guilty by reason of insanity. In response to this decision, Queen Victoria requested that the House of Lords debate the insanity issue. The resulting M'Naghten Rule stated that a defense of insanity required proof that the accused party, at the time of committing the act, was laboring under defective reason and did not know the nature and quality of his act—or, if he did know what he was doing, did not know it was wrong.[7]

Soon to become the litmus test for determining insanity in both English and American courts, the M'Naghten Rule was narrowly constructed. By focusing on cognitive understanding of right and wrong, it ignored important features of the personality that involve one's emotions and the ability to control them. Thus, many jurisdictions added to it the "irresistible impulse" rule, allowing the spontaneous lack of control over emotional responses to serve as a defense of a criminal act. Covering cases in which the individual had cognitive capacity but could not control actions at a particular moment, the irresistible impulse rule did not deal with unconscious influences that might affect a person's ability to conform his conduct or control his behavior. Nor did it deal with broader personality features that can affect emotional control and the exercise of free will.

The decision in the 1954 *Durham v. U.S.* case attempted to fill these gaps by excusing unlawful acts of homicide if they could be shown to be a "product" of a "mental disease or defect."[8] This expanded interpretation led to the greatly increased use of psychiatry and psychosocial information in the courts. Indeed, for a while courts interpreted the Durham test so broadly in their instruction to the jury that virtually any psychiatric information suggesting a linkage between an offender's personality and the unlawful act justified acquittal.

Durham did not limit mental disease or defect to very serious disorders, nor did it require a close link between the disorder and the "product," or criminal action. The decision suggested that even the social and economic background of an offender could prevent full exercise of free will. Soon the court began to carve away at such expansive interpretations, and in *Carter v. U.S.* (1957) it limited the definition of the "product" of a mental disease to a strict relationship between the act and the offender's mental condition.[9] The offender had to prove that he would not have committed the act had he not been suffering from the condition. Subsequently, in *McDonald* (1962) the court limited mental disease and defect to an "abnormal condition of the mind which substantially affects mental or emotional processes and substantially affects behavior controls."[10]

In the 1972 *Brawner* decision,[11] the U.S. Court of Appeals overruled *Durham* and adopted a rule that had been generated by the American Law Institute in 1955. The ALI rule states that an individual is not responsible for criminal conduct if, as a result of a mental disease at the time of the act, he lacks "substantial capacity" to appreciate the wrongfulness of his conduct or "to conform his conduct" to the requirements of the law.[12] This rule had been gaining support in various states since the mid-1950s as an alternative to the M'Naghten/irresistible impulse test. Further narrowing the application of the insanity defense, the rule had been adopted by all of the federal courts and an increasing number of state courts. In addition, case holdings and legislative developments had already begun to limit the role of the psychiatrist in the courtroom. In *Washington v. U.S.* (1967), Judge David Bazelon of the U.S. Court of Appeals for the District of Columbia expressed his discontent with psychiatric labels and the "paucity of meaningful information presented to the jury." He held that psychiatrists as expert witnesses could not come to conclusions regarding the appropriateness of the insanity defense but should be limited to describing the mental condition of the offender, leaving it up to the jury to decide whether the offender's mental status related to the criminal conduct.[13]

But the ALI rule did not resolve the tensions between psychiatric understanding of human behavior—strongly deterministic and weighted toward exculpation—and society's concerns for stability, order, and safety. While considered a major improvement over *Durham*, the ALI rule neither limited the appropriateness of psychiatric opinion nor provided guidelines to the jury for assessing the validity of psychiatric testimony. It thus became the object of considerable controversy during the 1982 trial of John Hinckley for his assassination attempt on President Ronald Reagan.[14]

When Hinckley was judged not guilty by reason of insanity, the public was outraged. His commitment to a mental institution gave little comfort to those who insisted that the insanity defense had compromised their concerns for punishment and deterrence. The tensions were expressed in the conflicting evidence presented at the trial. In attempting to evaluate Hinckley's ability to assess reality and to conform his behavior, experts produced over a dozen diagnoses.[15] Experts testifying for the defense found Hinckley delusional about actress Jodie Foster; experts for the prosecution said he was merely infatuated with her. While both sides agreed that Hinckley had a mental disorder, they disagreed on the type of disorder and on whether it rendered him incapable of controlling his behavior or understanding that it was wrong. The psychiatrists for the prosecution testified that Hinckley was sane and those for the defense testified he was insane.[16]

The outcome of the trial rested partly on the requisite burden of proof. The prosecution had to prove that Hinckley was sane beyond a reasonable doubt, and that he did not have a mental disorder impairing his ability to conform his conduct or appreciate its wrongfulness at the time of the crime. The psychiatric experts relied on phenomenological assessments, performing the traditional mental status examination to assess Hinckley's emotional character and thinking process. Such expert evidence is opinion evidence.[17] With no biological information on which to base its conclusion, the prosecution, in the jury's view, failed to prove Hinckley's responsibility for his act.[18] The defense, however, introduced evidence from a computer-assisted brain scan showing that Hinckley's brain had "shrunk," according to the expert interpreting the data. The court reluctantly admitted this evidence, though there is in fact no proof of a link between mental illness and brain size.[19]

The adversary system often polarizes psychiatric opinion. The absence of definitive biological markers of insanity leaves room for broad disagreement. Differences in the philosophical perspectives of psychiatrists, in the procedures used by attorneys to select psychiatrists as expert witnesses, and in the data obtained by separate examinations contribute to the polarization so evident in the *Hinckley* case. They also lead to demands for "harder" data that would preclude biases and reduce disagreements.

In a survey assessing the public's reaction to the verdict in the *Hinckley* case, 40 percent of those who responded indicated they had essentially no confidence in the psychiatric testimony during the trial. An additional 20 percent indicated they had, at best, slight confidence in the testimony.[20] Phillip Resnick, a well-known forensic psychiatrist at Case Western Reserve University, points out that such skepticism about psychiatric testimony is not limited to the public. Judges and psychiatrists themselves are among the most severe critics of its scientific validity.[21]

Skeptical reactions to psychiatry have been long-standing and widespread. A survey in 1972 indicated that the public had significantly more confidence in general medicine than in psychiatry.[22] Other surveys suggest that psychiatrists are rated low in terms of honesty and ethical standards.[23] The attendant mistrust of "soft" psychiatric evidence has come at a time of increased demands on the legal system to balance the need for social order with protection of individual rights. These pressures encourage the admission of harder scientific evidence in the courts.

ASSURING EFFICIENCY: ADMINISTRATIVE STRAINS ON THE COURTS

The caseload of most state and federal courts has exploded in the last ten to twenty years, without a corresponding increase in the number of courts, judges, or court administrators to absorb the burden. The situation in the State of Washington provides a typical illustration of the pressures that confront appellate courts throughout the country. In 1974 the Washington Supreme Court had 497 filings. These included a variety of petitions, motions for discretionary review, actions against state officers, and disciplinary matters. In 1983 it had a total of 913 filings, an increase of approximately 84 percent. The justices of that court are currently writing between 1,300 and 1,800 opinions per year.[24]

The caseload in the Court of Appeals in Washington State was even more burdened in those years. During 1974 it handled over 1,500 filings including appeals, personal restraint petitions, and motions for discretionary review. By the mid-1980s its caseload had doubled, with more than 3,000 filings. Interestingly, there was a parallel discrepancy between the cases pending and those resolved during that period. In 1979 the Court of Appeals disposed of nearly as many cases as were pending. By the mid-1980s that ratio had decreased to approximately 78 percent, leaving many cases held over until the following year; the court simply could not assimilate a 10 percent average annual increase in the number of cases.[25]

The administrative overcrowding of the court system has focused attention on the time it takes to process cases. Delays can be due to a number of factors, including various economic or social incentives; strategic considerations on the part of the participants; the time the attorneys are willing to put into the case; and the complexity of the case, including the nature and range of infractions that need to be reviewed, the number of defendants, and the time needed to collect evidence.

Lengthier processing time results in increased expense for both sides in a dispute. Moreover, delays can significantly affect the outcome of a case. All sides in a criminal case need time to obtain the necessary facts, as well as to organize the testimony of witnesses. But a long processing time is especially detrimental to the interests of the state because the longer it takes to close a criminal prosecution, the less likely it is to convict an offender.[26] Witnesses are likely to be less available and to forget details that can be crucial to the final decision. The economic costs of such administrative difficulties encourage the use of technologies to provide definitive evidence that will both speed up the processing time and facilitate more efficient resolution of criminal cases.

New diagnostic procedures can help assure speed and accuracy in the resolution of cases by reducing the amount of expert opinion testimony required. By resolving questions of fact, accurate scientific tests can eliminate the need for large quantities of contradictory evidence and encourage early settlement.* When

*It could also be argued that improved testing, by increasing the amount of evidence available for dispute, may add opportunities for litigation.

the courts are faced with demands to sort out complex questions of fact and law, diagnostic tests can simplify and resolve fact finding, narrow the decision-making process, and reduce the time it takes to settle a case either out of court or in a trial.

Existing Strategies

The courts have developed several strategies during the past fifteen to twenty years in response to such pressures. They have tried to define an increasing range of problems in medical or scientific terms. And they have tried to buttress their judgments through scientific approaches, in effect extending the numbers and types of cases amenable to technical resolution.

MEDICALIZATION

"The law," says legal analyst Michael Moore, "is not neutral about the various ways people may be explained."[27] Lawyers have had a long-standing interest in finding biologically determined causes for the behavior of their defendants. For example, in 1924 Clarence Darrow tried to explain the Leopold/Loeb murder of Baby Franks in terms of Loeb's endocrinological problems. And the phenomenological explanations of criminal behavior advanced by Cesare Lombroso in the nineteenth century enjoyed a long popularity.[28] However, until recently the parties to a legal dispute, and occasionally judges, looked mainly for psychological explanations of aberrant behavior—for example, calling on experts in early childhood development to characterize a defendant and speculate on his intentions. Today, lawyers and courts are increasingly seeking biological, organic causes to provide plausible explanations of abnormal behavior. By medicalizing deviance, the law finds "objective" explanations to defend and prosecute in the court.

A series of cases over the past decade suggest the trend toward medicalizing deviance through recourse to current biological theory and diagnostic tests. In each of these cases biological explanations have been brought to bear on the insanity defense. In 1979 a New York court judged a policeman who shot a fifteen-year-old male not guilty because he had a psychosis associated with epilepsy and hysterical dissociation under stress.[29] During the trial the court accepted, and the jury affirmed, the association of temporal lobe seizures with uncontrolled violence, even though neither the existence of seizures nor their association with violent behavior had been clinically or scientifically established in this specific case. After the trial the offender was committed to a mental health facility for a thirty-day evaluation, during which time an EEG revealed diffusely abnormal brain waves, though no specific indication of temporal lobe or psychomotor epilepsy. Following the evaluation the Commissioner of Health's petition for release of the offender was granted. The State appealed the release to the Supreme Court of the Appellate Division, which reviewed the lower court decision. On appeal, experts disagreed about the mental status of the defendant: some testified that he was not mentally ill; others agreed with the initial diagnosis. The psychiatrist interpreting the EEG claimed that the test allowed the possibility of temporal lobe epilepsy. However, the New York Court of Appeals affirmed the decision that the defendant was no longer suffering from mental illness and reinstated his release.

A defendant in a 1982 Colorado case was acquitted after a neurologist testified that the neurotransmitters in his brain "malfunctioned."[30] His diagnosis was organic brain syndrome that left him with poor impulse control. Three years later a man in Florida was found not guilty by reason of insanity for murder and armed robbery when he was diagnosed as having organic brain damage caused by fumes from a solvent, aggravated by drug use.[31] The same year another Florida jury acquitted a defendant of murder by reason of insanity, but found him guilty of armed robbery the night prior to the murder.[32] He was judged as having been sane on the night of the robbery, but the defense experts, who diagnosed schizophrenia plus organic brain syndrome aggravated by chronic drug abuse, convinced the jury that the murder of the police officer chasing him the following night was an act of insanity.

In such cases, medical experts become part of the adversarial system, providing their opinions concerning the defendant's mental state which will be measured against the legal standard for insanity. Experts for each side battle it out under the guidance of their attorneys and the challenge from the opposing counsel. The jury, though often unable to evaluate the disputed evidence, must weigh increasingly complex medical testimony to arrive at the truth. Their task is complicated by the tendency of medical experts to give opinions on legal as well as technical matters.

After John Hinckley's trial and the public's response to the verdict, Congress wanted to eliminate "the confusing spectacle of expert witnesses."[33] Both the American Bar Association (ABA) and the American Psychiatric Association (APA) recommended comparable reforms in the insanity defense. The ABA wanted to focus on "legitimate and more objective psychiatric factors." Seeking more objective measures, it recommended striking the volitional arm of the insanity test, finding no scientific basis for measuring a person's capacity or lack of capacity for self-control.[34] It endorsed the cognitive "right from wrong" test, arguing that if because of a mental defect the individual did not realize his conduct was wrong, he could not be held responsible.

The APA, noting the disagreement among psychiatrists over the concept of volition, cited the 1979 *Addington* case, in which the court found that psychiatric evidence was not sufficiently clear-cut to prove or disapprove many legal issues.[35] In this case,

Supreme Court Justice Warren Burger stated that "Psychiatric diagnosis . . . is to a large extent based on medical impressions drawn from subjective analysis and filtered through the experience of the diagnostician."[36] In a subsequent decision, *Ake v. Oklahoma*, Supreme Court Justice Thurgood Marshall explicitly stated the limits of psychiatric testimony: "Psychiatry is not, however, an exact science, and psychiatrists disagree widely and frequently on what constitutes mental illness, on the appropriate diagnosis to be attached to given behavior and symptoms, on cure and treatment, and on likelihood of future dangerousness. Perhaps because there often is no single, accurate psychiatric conclusion on legal insanity in a given case, juries remain the primary fact finders on this issue."[37]

The discrepancies in psychiatric opinion had been dramatically revealed by the multiple diagnoses of Hinckley's disorder from both prosecution and defense experts. To minimize such problems, the APA recommended limiting psychiatric testimony to the diagnosis of the defendant's psychological state of mind. The psychiatrist was "to do psychiatry" and avoid testifying on whether the defendant was *legally* sane at the time of his act or whether he had the requisite legal state of mind. Such "ultimate issues" should be left to the jury.[38]

The Insanity Defense Reform Act of 1984 codified the positions of both the American Psychiatric Association and the American Bar Association for insanity defendants tried in federal courts.[39] Modifications to the insanity defense placed limits on expert opinion, disallowing opinion evidence on the legal issues of a case.[40] Under the revised Federal Rule of Evidence 704(b), experts must testify about the defendant's "medical" mental condition, not his legal mental state. They must impartially decipher symptoms and present them credibly and persuasively.[41]

By limiting opinion evidence on anything other than the defendant's mental state, the reforms of the insanity defense have gradually weakened the traditional forms of psychiatric testimony. They have, in effect, institutionalized medicalization by emphasizing the importance of "hard" biological evidence rather than expert opinion.

THE SEARCH FOR SCIENTIFIC EVIDENCE

The legal system has rules governing the admissibility of scientific evidence in the courts. It is the judge who determines admissibility, usually by weighing the probative value of evidence against how much it might prejudice the defendant. In their conservative tradition, the courts are reluctant to admit new scientific evidence. It must first meet the Frye "novel science" test, which requires that the theory be generally accepted in the scientific community as statistically verifiable and demonstratively provable.[42] The criteria for judging the scientific acceptability of a test are loosely defined (except in the state of Michigan, which has established specific standards). Thus, the Frye test is repeatedly applied whenever attorneys seek to bring in novel scientific evidence. But once admitted, scientific or medical evidence is highly persuasive; it tends to be accepted by juries as the truth and is not easily refuted by "softer" opinions.

Defendants have used novel "scientific" defenses, such as the Twinkie defense (lack of control caused by imbalance in body sugar) or the post-Vietnam syndrome (lack of control caused by too much stress) with varying success. Defenses that fail often have no biologically provable foundation, no abnormality that excuses behavior. This was the case in the efforts to use premenstrual syndrome (PMS) as a defense for criminal behavior.

PMS has been advanced as a type of temporary insanity, that is, a mitigating circumstance.[43] The medical theory is that in some women the biological and chemical changes accompanying the menstrual cycle can be responsible for irrational behavior, including criminal acts. But as a defense for criminal behavior,

PMS has not worked in American courts because of limited scientific evidence. As many as 90 percent of all women may exhibit at some time one or more of the approximately 150 symptoms of PMS, and 40 percent have had disabling PMS symptoms.[44] But fifty years of clinical study have provided neither comprehensive diagnosis nor conclusive medical tests to define the syndrome.[45] PMS has been called a "dynamic interaction of neurobiological, psychological and social elements."[46] Some doubt whether it is a "disease" at all.[47] And even if there were evidence associating PMS with a specific biological disorder, using it as a defense in the courts would be strongly resisted in the United States where feminists fear that it would result in sexual discrimination and the stereotyping of women as "endocrinological cripples."[48] They argued that if a biological cause were found, it could revive old myths about women's capacity to function during PMS days.

In an English court, however, charges against two women were reduced from murder to manslaughter because of PMS. In Canada, PMS has been used to mitigate a sentence, and France recognizes PMS as a legal insanity defense.[49] But the United States has had no successful PMS defenses, for neither the psychiatric community nor the courts have accepted PMS as a mental disorder.[50]

Though medical experts are limited in the scope of their testimony, demands for objective scientific evidence have in fact expanded their role. The courts seek scientific and medical information as the basis for decisions in a wide range of legal matters, including civil cases involving issues such as paternity, child custody, testamentary capacity, competence to enter into a contractual relationship, and responsibility for tort actions.

In tort or negligence law, the courts are demanding scientific facts as the basis for deciding liability. Developments in the natural sciences have long provided the facts to buttress negligence claims. Scientific knowledge about the stresses that may compromise metals under adverse conditions has aided the understanding of events such as the Challenger III space shuttle disaster. Understanding the physics of stress has frequently contributed to tortious claims concerning defects in the construction of bridges, high-rise apartment buildings, and other public facilities. Similarly, new diagnostic technologies can provide precise information about certain types of human injury. For example, the introduction of fetal monitoring during the last stages of pregnancy allows physicians to observe previously undetectable changes in the status of the fetus. This diagnostic technology has enhanced the ability to predict and avoid difficulties during the birth process. It can also reveal the reasons for fetal injuries that may have been caused by professional negligence.

Scientific strategies are also employed in family law to determine paternity. In the past, the courts relied on the history of family relationships and superficial characteristics of individuals with nonmatching blood types to determine paternity. But over the past decade, such determinations have routinely been made on the basis of biological evidence. As increasingly sophisticated and accurate techniques, such as human leukocyte antigen (HLA) blood tests, became available in the early 1980s, the courts quickly adopted them. The HLA blood test, which involves tissue typing of white blood cells, determines paternity with a high degree of accuracy. Earlier blood-typing tests could only exclude a particular man as the father of a child if his blood type did not match the child's. If the cells matched, that did not necessarily establish paternity. HLA, however, provides direct evidence of paternity.[51]*

In a 1982 case, a Kansas court found the HLA test to be 99.6 percent accurate. Vigorous cross-examination of the expert who had conducted the test, and verification of rigorous laboratory conditions emphasized its conclusiveness.[52] However, in a 1983

*Admissibility may be limited by statute in states that allow blood test results only to *exclude* paternity. In those states HLA is not admitted because it *proves* paternity.

case, the Kansas court found HLA evidence inadequate, unfair, and prejudicial to the defendant. In that case, an expert testified that HLA results are only 75 percent accurate, while the judge required at least 90 percent accuracy before he would admit the evidence.[53] Today the statutes of forty-seven states admit genetic test evidence to prove paternity, for such tests do not depend on expert witness testimony to assess the validity of their results.[54] The data speak for themselves. Based on neither numerical probabilities nor scientific theory but on empirical data, they are usually admissible evidence.

Those technologies subject to conflicting interpretation are less likely to be admissible in the courts. Polygraph examinations can assess the truthfulness of an individual's response to a technician's questions with a potential accuracy of 90 percent. In most individuals, lying creates psychological stress; the sympathetic nervous system responds involuntarily to that stress and produces physiological responses that can be measured by a polygraph. These measurements demand skilled interpretation under carefully controlled conditions.[55]

Problems of accuracy arise when these tests are not carefully administered and interpreted, and thus the polygraph has had a consistently hostile reception in the courts.[56] Initially the Frye test excluded lie detector evidence because it was not based on generally accepted scientific principles. Today, courts are reluctant to admit it because, unlike the results of HLA tests, polygraph evidence is opinion; the results do not speak for themselves. A person skilled in the "art and science" of interpreting polygraph results gives an opinion on the veracity of the defendant's response. Moreover, interpretation is complicated by the limits of the test. Results may be influenced by drugs, and some individuals are able to countermand the test.[57]

The courts have been cautious in their acceptance of scientific information. The trend, however, has been to rely increasingly on scientific and technical evidence obtained from diagnostic technologies, especially as their results appear more objective and, therefore, more compelling than the evidence provided by expert opinion.

New Technologies and Future Developments

A number of new scientific and technological developments based on genetic and neurobiological research are providing forensic specialists, courts, and juries with a way of determining the truth in a range of criminal and civil cases.[58] DNA fingerprinting techniques are currently being used as evidence in the courts to establish the identification of suspects in criminal cases. The courts are also beginning to admit diagnostic techniques that correlate violent behavior with brain abnormalities and thereby provide evidence of the inability to control behavior or to assume responsibility. Research in neurobiology and the initial use of imaging techniques in the courts suggest the direction of future applications both in defining insanity and in predicting behavior.

In 1987 a court in Orlando, Florida, accepted DNA fingerprinting in establishing the identity of a rapist.[59] The prosecution introduced into evidence a DNA fingerprint derived from semen stains on the victim and a DNA fingerprint from a blood sample of the alleged rapist. The technique was used to prove that the DNA of the blood matched that of the semen stains. Scientists are able to examine blood stains, semen stains, and hair roots to determine the specific DNA genetic fingerprint that is believed to be unique for each individual. There continues to be some debate about appropriate testing procedures, and concerns about quality control in the collection, handling, and storage of samples. But most scientists agree that there is almost no possibility that two individuals will share the same pattern of DNA fragments (unless they are identical twins). In evidentiary terms, DNA prints offer conclusive scientific proof. Convinced by the scientific consensus about its reliability, more and more courts are admitting this new technique as evidence.

Because the DNA fingerprinting test is so accurate, its probative value is high. Its use, in effect, obviates counter expert opinion. Once the DNA evidence was entered in the Florida case, the defense did not call in a single witness to counter it. Moreover, once the test is admitted, a jury is unlikely to question its accuracy and truthfulness. Expressing an unquestioning faith in science, a juror in a New York State rape case explained his decision: "You can't really argue with science."[60]

The highly compelling nature of such scientific evidence, however, holds some danger. Because of its statistical accuracy, the DNA test can undermine the presumption of innocence that must guide a jury's deliberation, and thereby eliminate its discretionary role. A jury must consider a very wide range of factors influencing the behavior of a defendant. Given the endemic faith in statistics, such factors could be submerged by the weight of the evidence about one particular aspect of the case, creating a situation of unfair prejudice. This problem was addressed by a series of cases in Minnesota prior to the availability of DNA fingerprinting, in which experts used blood and semen analyses to identify rapists. Despite the accuracy of these tests, the courts excluded the statistical evidence on grounds that it would undermine the presumption of innocence and unfairly prejudice the jury.* Expressing concern about the power of the statistical infor-

*The Minnesota cases directly addressed the introduction of statistical information and its potential for unfair prejudice in criminal trials. *State v. Carlson*, 267 N.W.2d 170 (Minn. 1978) involved the brutal murder of a twelve-year old girl. Comparing blood samples and pubic hairs, experts provided statistical data to identify the suspected criminal. The court, however, refused to admit the information, questioning "the psychological impact of the suggestion of mathematical precision" and its effect on the jury's role. A subsequent case, *State v. Boyd*, 331 N.W.2d 480 (S.Ct. Minn. 1983), involved an allegation that the defendant had engaged in criminal sexual conduct with a fourteen-year-old girl who became pregnant. Again the court did not allow testimony based on blood tests indicating the probability, 99.911 percent, that the defendant was the father. According to the

mation and the "psychological impact of the suggestion of mathematical precision" on the jury, the judge in *State v. Carlson* (1978) stated that "testimony expressing opinions or conclusions in terms of statistical probabilities can make the uncertainty seem all but proven."[61]

Addressing these concerns the defense attorney in the Florida case worried that the admission of evidence from DNA fingerprinting would invade the right of the jury to make decisions, and might even introduce the issue of whether juries are needed at all when such evidence is available. Nevertheless, following the Florida precedent, courts have used DNA fingerprinting to improve the accuracy of identification. By June 1990, DNA analysis had been admitted into evidence in at least 185 cases in thirty-eight states, though many questions remain about the reliability of test results. Commercial firms engaged in forensic DNA work are receiving hundreds of requests. The FBI is training state forensic scientists to use the technology and setting up DNA data banks similar to fingerprint files.[62]

As an accurate means of identification, DNA fingerprinting is being used in other contexts as well. In the United Kingdom, immigration authorities are taking DNA fingerprints from blood samples of immigrants seeking entry into the country. The officers are attempting to establish the paternity of the male in each group that is seeking admission, because they prefer to allow only families to immigrate.[63] U.S. immigration authorities are among those requesting fingerprint tests from the commercial laboratories. The use of these tests for screening suggests their potential effect on the legal status of families and the rights of fathers. Offi-

---

judge, "the jury should be made to understand that the frequency figure does not in any sense measure the probability of defendant's innocence." The third Minnesota case, *State v. Joon Kyu Kim*, 398 N.W.2d 544 (S.Ct. Minn. 1987), involved a defendant who was convicted of third- and fourth-degree criminal sexual assault. The State Appeals Court excluded statistical evidence based on blood and semen analysis, arguing that such measures of probability would undermine the presumption of innocence. In all three cases, the concern was not the nature of the evidence but its impact on the jury and the result of unfair prejudice.

cials in the U.K. have also passed legislation that would allow the police to take mouth swabs from suspected terrorists in Northern Ireland without consent.* The swabs would provide genetic evidence for "profiling purposes," and the fingerprints could later be tested against traces of tissue found elsewhere during the course of an investigation.[64]

Tests intended to document the biological basis of criminal behavior as a support for the insanity plea have found less acceptance in the courts. In the late 1960s and early 1970s medical researchers studying prisoners found an apparent correlation between violent behavior in some males and the existence of an extra Y chromosome in their genetic makeup.[65] Researchers asked whether the XYY chromosomal abnormality predisposed an individual to violence, and, if it did, whether it could predict violent behavior. Studies of the incidence and prevalence of the XYY syndrome, focusing on hospital and prison populations, had substantial methodological problems.[66] While the prisoners did show a prevalence of the syndrome, the evidence did not support a classification of XYY males as aggressive, nor did it allow prediction. Not all XYY males are in fact violent. According to one source, the only generalization to be drawn from the XYY anomaly is an atypical socialization of excessively tall males.[67] Nevertheless, some attorneys attempted to use the XYY syndrome as evidence of a mental defect that deprived defendants of free will and, therefore, responsibility.

The XYY studies were part of a long history of unsuccessful efforts to correlate violent behavior with brain or genetic abnormalities. These efforts, initially spurred by animal research during the 1920s and 1930s, associated functional problems of various brain areas with violent reactions. In the early 1970s, an aborted project involving sexual psychopaths sought to assess the effect of destroying a small portion of the brain on the control of

*The swabs are taken from the lips and can therefore eliminate both the obstacle of clenched jaws and the need to break teeth.

psychosexual violence. Researchers viewed the destruction of brain tissue as a way both to test the source of violence and to effect a psychosurgical therapy.[68] The study was blocked by the legal and ethical issues associated with informed consent in a prison population, and by the limited data available to back up theoretical assumptions about the relationship between the portion of the brain to be destroyed and violent behavior.*

Today, less invasive and more sophisticated techniques are used to observe brain abnormalities. Positron Emission Tomography (PET) scans of violent patients have linked their deviant behavior to specific abnormalities in the brain.[69] In 1987 PET was used in a pilot study of the biological bases of violent behavior. Four patients who had engaged in repetitive, nonpurposive violent acts were tested. The PET images of two of these patients showed compromised function of the frontal cortex (related to cognitive functions) and the right temporal area (associated with emotions). These patients had a history of repeated acts of violence and showed no comprehension of the ethical implications of their behavior. The other two patients showed deficits only in the temporal cortex. They had engaged in violent behavior characterized by impulsivity and lack of control. Interpreted against the background of animal research, these results suggest that the patients' behavior followed specific neurological impairments, and that specific abnormalities in the brain could be used to anticipate outbursts of rage, incapacity to control violent impulses, or lack of moral comprehension of the nature of an act.[70] Similar studies are under way to examine the brain metabolism of sexual psychopaths who engage in violent behavior.[71]

PET has the possibility of replacing guesswork in the psychi-

*The study, which was to be conducted by the Lafayette Clinic of Wayne State Medical Center, involved examining the comparative effectiveness of amygdalectomies (destruction of part of the limbic system involved with the generation of emotions) and antitestosterone treatments for the control of psychosexual behavior. The recidivists who were the subjects of this study were to be divided into two groups in order to compare the two treatments involved.

atric evaluation of case histories. It can also provide substantive evidence to support predictions of violent behavior. For example, defects of the prefrontal cortex, implicated in disturbances of cognitive functioning, may be used to suggest whether a defendant is likely to continue to be dangerous and to lack a sense of moral responsibility for his actions. This has important implications for assessing the rehabilitation potential of convicted criminals in cases such as *Barefoot v. Estelle* (see page 133).

While PET research is only beginning to examine such problems, professional observers are optimistic about its application. In 1985, the authors of a book on the insanity defense claimed that "a defect in the brain can be diagnosed now by CAT scan, PET scan, NMR* tests, EEG readings, evoked potential readings, or hair analysis. We now have direct scientific evidence as to the state of the defendant's brain at the time of the examination, and we need not rely on verbal reports and questions such as, 'Did the defendant know right from wrong?' "[72]

In fact, evidence from PET scans has been admitted in California in criminal cases to judge the sanity of murderers and in tort cases to evaluate head injuries. In a 1985 murder case, a California court allowed an expert to testify on the results of a PET scan made on a defendant who had murdered four members of his family. During the trial psychiatrists had disagreed about the offender's mental condition, but then the PET scan revealed severe hypofrontality (decreased metabolism of frontal lobes)—a condition that is consistent with schizophrenia. Initially, the judge did not allow the expert to present the images from the scan because he was concerned that it would prejudice the jury. However, when the judge found that the images were consistent with the expert testimony diagnosing schizophrenia, he admitted six of them at the second stage of the trial, when it is determined whether the defendant was insane at the time of the act. The jury, however, found the de-

*Nuclear Magnetic Resonance. The name of these tests was changed, for political reasons, to Magnetic Resonance Imaging.

fendant guilty. Although the PET evidence helped establish that the defendant was mentally ill, the jury did not find that his illness met California's legal definition of "insanity." The evidence did, however, influence his sentence; the jury gave the defendant life imprisonment rather than the death penalty.[73]

In another case, Walter Hodge, who had been clubbed on the head three times by a policeman, brought suit for damages, but his lawyer postponed filing the case and the statute of limitations expired. Hodge then brought a negligence action against the lawyer. To show that the original case had merits, he introduced a PET scan that revealed brain abnormalities consistent with a head injury.* The court admitted the PET evidence and Hodge won a settlement of $2,735,000 for legal malpractice.[74] A related, and far less costly brain imaging technology, BEAM, has also been frequently admitted by courts into evidence in both criminal and civil cases.

The redefinition of criminal behavior resulting from biologically caused deviations comes at a time when reforms of the insanity defense are challenging the role of "soft" psychiatric expertise in the courts. Indeed, the growing practice of relating psychopathic behavior to specific malfunctions of the brain is leading to new approaches to problems of criminal insanity. According to some analysts: "We have scientific means of establishing disorders of the brain, and we can replace the concept of insanity with neurological concepts of diseases of the brain. Such disorders as episodic dysfunction, premenstrual syndrome, post-traumatic stress disorder, hypoglycemia, XYY Syndrome, and psychopathology must be given some legal standing. Since the neurological conditions do not constitute insanity or mental illness, new approaches must be found and new scientific definitions of brain disorders established. We must decide if such biological conditions are to be treated as illness or punished as crimes."[76]

*The *Los Angeles Times* reported on Dr. Monte S. Buchsbaum's testimony that a PET scan performed on Hodge revealed that he had lost the use of 10 percent of the frontal lobe portion of his brain, just below the two-inch scar.[75]

Biological approaches to insanity are changing the nature of acceptable evidence and the configuration of legal expertise, ultimately giving far greater power to the technical expert and to those who can pay for innovative scientific techniques.[77] As the disciplines of neurophysiology, biochemistry, and genetics find biological explanations for deviant behavior, and new techniques provide the data base for associating brain dysfunction with aggression and violence, new forms of legal evidence are being admitted to the courts. If an imaging technique can watch the brain function, it provides our legal institution with a powerful weapon. If the technology can make an abnormality objectively visible, it can help society to categorize behavior. Evidence that can be tangibly observed on a computer printout, read in a PET scan, traced in a computerized electroencephalogram, or visualized in an X-ray assumes the aura of truth and gives our legal institutions a greater power to screen, to stigmatize, and to control.

CHAPTER 8

Social Control Through Biological Tests

RECENT biomedical discoveries have redefined a broad spectrum of diseases and behaviors as biologically determined, diagnosable, and predictable. Advances in genetics and the neurosciences have provided powerful instruments for biological profiling which can predict the physical disabilities and behavioral abnormalities that an individual might develop later in life. Tests are being developed that are efficient, inexpensive, and unobtrusive. It will be simple to test every newborn child. DNA storage techniques allow genetic material to be frozen indefinitely. Computerized DNA "banking" allows instantaneous retrieval of genetic information. It is quite feasible to establish national data banks to store information about a person's parentage and predisposition to disease. Indeed, some biotechnology firms are predicting that most people will have their genetic profile on record by the year 2000.[1]

The medical benefits of anticipating genetic disease, the social benefits of records that would facilitate control of "criminal elements," and the economic benefits of having data for rational planning serve as powerful incentives for developing a data bank program. Such a program would also fit with what we have called the actuarial mind-set—the faith in facts, the need for economic efficiency, and the tendency to reduce complex problems to manageable and measurable dimensions. Moreover, a testing and data bank program would meet the powerful economic pressures for conforming individuals to institutional goals.

We have examined the social implications of the increasing preoccupation with testing in American society, focusing especially on the diagnostic techniques emerging from recent advances in genetics and the neurosciences. We have argued that certain cultural conditions legitimate the proliferation of tests: the faith in statistical information and the tendency toward biological reductionism in explaining social or behavioral problems. We have suggested how diagnostic techniques are being employed and can be expected to be employed in various institutions: in medical settings for purposes of planning and insurance, in the workplace, in the schools, and in the courts. In each of these institutions we explored how tests are routinely used as tools of mediation, by defining competence or responsibility, and as gatekeepers by controlling access and defining the boundaries of acceptable behavior in ways that support existing practices and policies. In this context we suggested the growing interest in sophisticated diagnostic techniques that can help organizations more accurately predict the future physical and behavioral characteristics of their clients.

Institutions must operate efficiently, controlling their workers, students, or patients in order to maintain economic viability. In some cases social controls are explicit, exercised through force; but more often institutions seek to control their constituents less by force than by symbolic manipulation.[2] Sanctioned by scientific authority and implemented by medical professionals, biological tests are an effective means of such manipulation, for they imply that institutional decisions are implemented for the good of the individual. They are therefore a powerful tool in defining and shaping individual choices in ways that conform to institutional values.

The use of tests to implement conformity is, for most organizations, economically rational. And frequently, convincing individuals to conform to institutional values is socially desirable as well. A society must, after all, control those who break the law, and organizations cannot function in a state of anarchy. But testing can also insulate organizations from change, increasing their rigidity and enhancing institutional control at the cost of individual rights.

Institutional Conformity and the Use of Tests

"Institutions bestow sameness," says anthropologist Mary Douglas. They "trim the body's shape to their conventions."[3] Anthropologists have long described the way cultures employ nature to support the ongoing social system. In our society, we call upon nature by using biological tests to assure that individuals conform to institutional values. Guided by the assumption that conformity will enhance efficiency and further their primary goals, schools, law enforcement agencies, employers, health care providers, and third-party payers use testing as a means of social control.

Testing enhances conformity in several ways. Biological tests can redefine institutionally problematic behavior as a problem of the individual, placing blame in ways that reduce public accountability and protect routine institutional practices. Viewed as scientific, biological tests give organizations a credible means to deal with failures without threatening their basic values or disrupting existing programs. When public schools face pressures for accountability from government and advocacy groups, it is convenient to explain learning difficulties or behavior problems in terms of individual disabilities. Biological tests redefine such educational failures as problems located in the student's brain. They help to remove blame from schools and other social influences. They justify medication as a means to assure conformity. And they provide diagnostic labels to satisfy the public's expectations that have given schools considerable responsibility to manage a broad range of social problems. Similarly, biological tests that identify the susceptibility of particular workers to toxic exposure may reduce a company's burden of responsibility to provide safe working conditions. Justified in the first instance as a way to protect worker health, such tests can be used to avoid costly changes in the workplace environment. It is the employee who assumes responsibility and is made to conform.

The predictive capacity of biological tests is especially useful to organizations in facilitating efficient long-term planning. Companies, as insurers, are reluctant to support those whose life-style or genetics may predispose them to a future illness. Employers routinely require pre-employment medical exams, from psychological tests for future executives to lower back X-rays for construction workers, from tests for drug use to screens for AIDS. In the context of growing economic competition, screening techniques that identify those predisposed to genetic disease can become a cost-effective way to control absenteeism, to reduce compensation claims, and to avoid future medical costs. Similarly, biological tests provide patient profiles for health care providers, helping them to control access to certain medical facilities, to conform to the reimbursement constraints of third-party payers, and to plan for future demands on the health care system.

Biological tests, perceived as scientifically objective, further enhance conformity by justifying difficult decisions about the exclusion of those who fail to comply with social or institutional values, or who threaten the economic need for efficiency. Faced with regulatory pressures and costly absenteeism, chemical companies have used screening tests to justify the exclusion of employees suspected of being genetically susceptible to chemicals. And because most employers provide insurance, the biological status of an employee, indicated by tests, may also be grounds for exclusion.

Tests may legitimate controversial decisions about the disposition of those who will not or cannot conform. In the hospital, tests can define an uncooperative patient as biologically incompetent, unable to make autonomous decisions and in need of paternalistic control. In the school, tests can justify decisions about the disposition of students and their access to special classes. And in the courts, judges are increasingly predisposed to admit hard evidence in order to sort out conflicting psychiatric opinion in decisions about the responsibility and culpability of defendants. Testing enhances the ability of these institutions to function more efficiently by facilitating the identification and disposition of those who may disrupt their goals.

Professionals as Institutional Agents

The power of the test to enhance institutional conformity relies on active professional support. Traditional professional relationships have changed in recent years, reflecting the growing medicalization of social and behavioral problems. Medical professionals are increasingly employed by schools, corporations, insurance companies, and the parties involved in legal disputes. These organizations depend on medical professionals to administer and interpret tests. Thus the increase in testing and the associated belief in the biological causation of disease and behavior have expanded the role of the medical expert in nonclinical institutions. Indeed, new diagnostic technologies have created many new professional jurisdictions.[4]

Many experts serve, in effect, as "company doctors" and are necessarily judged by standards enforced by their occupational association. If the priorities of their patrons conflict with their professional loyalties to their patients, they may be placed in the role of double agent.[5] For example, a physician employed by a school is beholden to his employer, but still subject to the professional ethic of primary loyalty to the patient. Responsible for administering and interpreting tests, the school physician may be called upon to make decisions about medication in the interest of classroom conformity rather than in the interests of the child.

Sophisticated biological tests are attractive to professionals who are faced with conflicting pressures, for tests can provide objective and therefore convincing data to back up difficult decisions. Such data can avoid social or ethical problems by redefining them in technical terms, but they do not resolve underlying questions of fairness or professional obligation. A PET image that comports statistically with a certain form of deviance and, therefore, suggests a specific treatment allows the professional to avoid examining that individual's life circumstances. A test that defines a child as hyperactive can shift the physician's focus away from other influences on behavior, such as a disturbed teacher or an abusive home environment. In effect, biological data encourages the domination of physical over social and ethical considerations.

In situations where experts face conflicting loyalties, the use of diagnostic tests can exacerbate professional dilemmas. When the competence of a patient is in question, a hospital-employed physician may be put in the position of interpreting tests to serve the priorities of the hospital system as well as the needs of the patients. Similarly, the availability of tests compounds the ambiguities in the role of those professionals who testify in court. Forensic psychiatrists must seek data to support the party who has hired them. The conflicts of interest inherent in this position existed long before the introduction of new technologies; but the use of biological tests adds to the problem. Opposing parties in a legal dispute may have unequal access to knowledgeable experts and to the resources necessary to employ novel and costly scientific techniques. Indeed, the high cost of such technical expertise can undermine adversarial norms. Once a test is presented to a jury as statistically valid, the burden of challenging its relevance becomes exceedingly difficult.

While the use of diagnostic testing has expanded professional roles, most institutions still rely on the information provided by primary-care physicians. This too is problematic. Many physicians lack the experience to deal with the probabilistic results of tests and their complex social implications. And testing procedures can limit a physician's ability to exercise independent judgment. Implementing a medical technology like PET involves a research team that includes neurologists, psychiatrists, chemists, laboratory assistants, and medical technicians from the hospital and sometimes also the medical instrument company. Clearly, decision-making power is diffused. Those physicians directly responsible for the welfare of their patients are no longer in complete control of the information underlying their decisions.

Finally, testing can transform doctor-patient relationships. As technologies become more complex and provide better characterizations of physical and psychological disorders, medical professionals are relying more on test results than on the symptoms of the individual. Credibility—provided by tests—seems to prevail over truth, especially when clinicians seek neutral data in areas of potential conflict. However, favoring test results over observations as a source of diagnostic information can be problematic. Most biological tests reveal the markers associated with a condition or disease; but, in fact, the statistical relationships revealed by a test may have little significance in the clinical or social context. Moreover, reliance on tests can create a presumption of causation when, in fact, there is only correlation. The more physicians rely on tests, the less involved they become with the patient as a person, and the more they distance themselves from the person, the less they are able to assess the implications of professional advice for the life of the individual involved.

Biological Discrimination

In 1988 Dr. Paul Billings, a medical geneticist and director of Harvard Medical School's Clinic for Inherited Disease, placed several advertisements in journals and magazines requesting information from people who had experienced genetic discrimination.[6] He received a range of responses. Most came from people who had been denied insurance after they were diagnosed as predisposed to a genetic disease. A man with an excellent driving record could not renew his automobile insurance when the company found out he had a neurological disorder, Charcot Marie Tooth Disease (CMT), though the disease had been stable and nonprogressive for twenty years. An eight-year-old girl, who had been diagnosed at birth as having PKU, was ineligible for insurance under a group plan, though with proper diet she had developed into a normal and healthy child. A young man diagnosed as hemochromatotic (excessive iron), but stabilized for many years through a regimen of phlebotomies (blood letting) was denied life insurance, though his parents, similarly afflicted, had lived into their eighties.

The response to Billings's inquiry also included reports of employment discrimination: an asymptomatic person denied a job because he had CMT; a carrier of Gauchet's disease rejected from military service. One of the most striking responses came from a young couple who had been advised by a genetics counselor to adopt a child because the wife's father had Huntington's disease; she herself had only a 50 percent chance of getting the disease and passing it on to her child. To avoid transmitting the defective gene to the next generation, the couple agreed to adopt. But when the adoption agency discovered their reasons for seeking an adopted child, it refused to process their application. Since the woman might become ill at some future time, probably in her late forties or fifties, she was not eligible to be an adoptive parent.

This case highlights some of the problems of biological discrimination. The woman faces several choices: She may decide to reproduce and assume the risk of transmitting the disease. Or she may decide to become pregnant but to have amniocentesis and, if the test is positive, then abort the child. But if the test is positive, it would also reveal that she herself will eventually get the disease, and this is information that she may not want to know. Finally, if she wants her application for adoption accepted, she may have to take a genetic test, an option that many people from Huntington's families have rejected.

The rights that people have come to expect in a free society are mediated by social institutions—the workplace, the schools, the health care system, the courts—but the economic imperatives and administrative needs of these same institutions often bring them into conflict with the social and individual values of their constituents. Testing for the biological origins of disease can affect our concepts of social equality, justice, and privacy, and our ideas about choice and free will. Screening to identify the genetic basis of intelligent behavior or disease can affect access to education and the right to work. Imaging techniques that detect subtle deviations can change the definitions of criminal responsibility and individual competence. The ability to predict disease can determine access to medical care as well as appropriate therapeutic procedures.

The rapidity of the commercial development and marketing of tests has troublesome implications. As marketplace interests increase in importance and economic considerations prevail over scientific goals, the quality and accuracy of tests may decline. Moreover, the marketing of new techniques, spurred by the existing market for prenatal tests, threatens to bring commercial values into reproductive decisions, to encourage a view of children as commodities amenable to a kind of quality control through ever more sophisticated intervention methods.[7]

Testing is ultimately a form of labeling, of distinguishing among individuals and placing them within specific categories based on biological criteria. While scientific knowledge provides a rational basis for classification, it also opens the way to discrimination, stigmatization, and vulnerability to control. Once a person is labeled as having a specific condition, or even a predisposition to one, others respond accordingly, attributing subsequent behavior or physical manifestations to that condition. Once a child like Casey Jessop, described in chapter 6, is diagnosed as hyperkinetic, teachers will be likely to interpret his later behavior in terms of that label, even when it represents a reasonable response to social provocation.

The label itself may accurately identify a specific condition, but it will also create a more generalized response to the person. Those in authority are likely to ignore the demands and dismiss the concerns of an individual labeled as "ill." A woman diagnosed as predisposed to depression may find her employer, her physician, and her family consistently associating her concerns with that underlying personality defect. A dissatisfied employee, whose medical record shows a genetic predisposition to manic depression, can be categorized as crazy, his complaints delegitimized as a manifestation of the disease. He may not necessarily become manic-depressive. As discussed in chapter 3, most markers indicate vulnerability, and the expression of disease may be contingent on other genetic and environmental factors. Yet the very presence of the marker will shape expectations about his health and label him in ways that will have economic, social, and psychological repercussions.[8] In effect, latent conditions take on the characteristics of disease, though they may have minimal behavioral manifestations.

Diagnostic labeling can also intrude on privacy. Data banks today contain considerable personal information about a large number of people. In some states information on hospitalized mental patients is stored in state files. Genetic registries record birth defects. The use of DNA tests in the courts implies the need for a national DNA data bank. Indeed, the FBI is creating such a bank, containing the DNA fingerprints of suspected criminals. While the purpose is limited to criminal investigations, stored DNA specimens collected for one purpose can be used for another unless the samples are destroyed.

Current advances in genetics and the neurosciences offer increased possibilities of storing biological information. Mapping the human genome will eventually make it possible to register an individual's genetic composition at birth. One can easily envision families demanding information about their "genetic roots,"[9] or commercial firms selling genetic information to interested agencies, such as insurance companies or the police. Questions of access to test results will juxtapose the privacy of individuals against the interests of relatives, employers, or organizations. Do members of a family have a right to information about the biological status of their relatives? Should a physician have the right, or perhaps the obligation, to communicate critical information about genetic disease to family members who may be similarly afflicted? Should people seeking to adopt a child be able to probe the genetic history of children available for adoption in order to "shop" for an appropriate match?

If the interests in biological information are sufficiently compelling, the privacy of individuals may well be compromised. In the *Tarasoff* case in 1974, a California court ruled that information provided by a patient in confidence to a psychotherapist must be disclosed if that information reasonably suggests that the patient is likely to injure another party.[10] In this case a patient told his therapist of his desire to murder his girlfriend. Believing the sincerity of his patient's claims, the therapist had him detained by the police. But the man was released in seventy-two hours. Several weeks later he, indeed, killed his girlfriend. Subsequently the California Supreme Court ruled that therapists have a duty to protect identified third parties who might be injured by their patients. In this situation *protect* meant informing the potential victim of the patient's destructive intentions. The compelling social interest overrides the individual's right to privacy.

Although the *Tarasoff* decision dealt specifically with psychotherapeutic relationships, the principal of duty to protect third parties has been extended beyond that situation. Thus *Tarasoff*-type arguments have been used to support the testing of vulnerable populations for AIDS and to inform potentially injured third parties. It is argued that the compelling social interest to prevent the transmission of AIDS should override concern about individual privacy. Similarly, in some circumstances, data from biological tests could be viewed as sufficiently compelling to warrant informing family members or employers, with little regard for individual rights. An airline pilot found to have a family history of a disabling genetic condition, for example, might be forced to undergo genetic testing or lose his license to fly.

Beyond their effect on privacy, biological tests can constrain people's choices. Once defined as sick, labeled as abnormal, or presented with evidence predicting disease, a person has limited options. To date, questions of choice arise most often in the context of genetic counseling. As more and more diseases are defined as genetic in origin, the number of genetic-counseling clinics grows. Recently, for example, clinics have been organized specifically to counsel families at risk of passing on mental disorders to their children. Counseling is currently based on a nondirective philosophy, that is, the counselor is supposed to provide genetic information to prospective parents and to lay out their options without advising them on which one to choose. However, most people faced with the prospect of a defective child have, in fact, little choice. For their options are constrained by economic possibilities, social pressures, and therapeutic alternatives. Whether the nondirective philosophy guiding genetic counseling will change with the increasing certainty of diagnostic measures and growing cost constraints is open to question.

Predictive tests can also circumscribe the choices of employees, extending the boundaries of corporate jurisdiction into family and personal life. We have seen how fetal-protection policies have constrained the reproductive decisions of some working women. Corporate interests also extend to personal habits such as smoking, drug use, and exercise—all believed to bear on a worker's performance. Employers and company physicians now ask most job applicants whether they have any physical or psychological conditions that might impair their ability to work. It is not a big step to inquire about genetic predispositions. As the sophistication and scope of testing for genetic markers increase, corporations could also justify routine testing of prospective employees for their predisposition to late-onset conditions that might someday impair them or result in chronic disabling conditions that would place economic burdens on health insurance coverage paid by the companies. Would corporate resources such as special training programs be invested in those individuals with "bad" markers? Can the right to be employed depend on having the "right" genes? Companies insure workers' dependents as well. As genetic linkage tests become available to diagnose predisposition to disease, can companies use such tests to deny insuring employees or their families?

Tests may further intrude on individual autonomy when they

are used to define competency. Competency implies responsibility. Lawyers, of course, are well aware of the relationship between competency and responsibility, and will routinely seek to establish a defendant's incompetency at the time of a criminal act in order to excuse violent behavior. But if a patient who refuses treatment is categorized as incompetent and the label is supported by a neurological test, the results can undermine the patient's choices in favor of hospital policy. Similarly, if a diagnosis of attention deficit disorder labels a child incompetent in the context of the classroom, this can limit parental choice.

The significance of diagnostic information rests, of course, on how it will be used. In this book we have described how schools, employers, insurers, and law enforcement agencies can use the information from tests. But other groups also have strong interests in the genetic or neurobiological condition of the people in their domain. The department of motor vehicles, immigration authorities, adoption agencies, creditors, organ transplant registries, professional sports teams, sexual partners, the military—all may have reasons for wanting access to diagnostic information. In some situations, their interest is appropriate. It may be reasonable, for example, to test those whose health may directly endanger the lives of others. But testing, we have suggested, is open to many inappropriate—indeed, dangerous—uses.

Considering the rapid development of diagnostic capabilities, there has been little concurrent discussion about the potential employment of tests in nonmedical contexts, the question of who should have access to biological information, and the obligations of avoiding abuse.[11] Concerns, however, are beginning to emerge. The American Society for Human Genetics, for example, has proposed guidelines to prevent genetic discrimination. Members of the European Parliament have questioned the European Commission's plans to integrate research efforts on the mapping of the human genome. Critics in the Parliament have insisted that the social and ethical implications of a "predictive medicine program," the potential for misuse of scientific knowledge, and, in particular, its eugenic implications, be addressed prior to undertaking the research. As the representative from the West German Greens put it: "It is crucial to have a proper understanding at this stage of the hazards which may be involved and not get too euphoric about the research."[12] A special committee of the Parliament has proposed that clear legal agreements be established with individuals concerning the use of their DNA and their rights over research results, that the implications of genome research receive wide public discussion, and that scientists accumulate information about possible consequences of the research before they develop ways of applying it. To facilitate public assessment, the Parliament's legal affairs committee has proposed an international commission for the ethical, social, and political evaluation of human genome analysis.

Meanwhile, the use of tests is self-reinforcing. Though institutions employ tests to increase efficiency, the very concept of efficiency becomes transformed by the standards developed through the use of tests. If diagnosis serves to enhance institutional values, these values themselves can be shaped by diagnostic criteria. Thus, testing seems to take on ever increasing significance in the life of organizations, serving as the basis to conform individuals to values that are themselves changing in response to diagnostic refinements. Given this power of diagnosis and its implications for therapeutic intervention, is it too futuristic to speculate on possible developments in genetic engineering or pharmaceutical manipulation? We tend to think of such techniques in therapeutic terms. But the dominance of tests and the value placed on institutional conformity suggest another scenario. Could not such techniques also be used to shape individuals for future institutional roles?

Before new diagnostic technologies are fully developed, we believe it essential to assure that questions of social justice and individual rights remain open to discussion and debate. For without such debate, economic interests in efficiency are bound to eclipse other values. Very early consideration of the diagnostic potential of research in genetics and the neurosciences is essential, for applications are inevitable. Such consideration should include the potential use of biological tests in nonclinical settings. Technology assessment is a recent development in medical care. Critical questions about medical effectiveness are being asked for the first time, and the results are published for professional and consumer scrutiny. Such assessments should extend beyond the purely medical setting to include investigation of the adverse consequences of testing as well as its benefits, its infringement on individual rights as well as the accuracy and validity of diagnostic information.

As sophisticated tests become available to detect more and more subtle conditions, the guidelines for genetic counseling should be reconsidered. Many genetic conditions may never be burdensome or significantly disabling. Criteria must be developed for defining and assessing the seriousness of diagnosed conditions and the probability of disease and its disabling consequences.

Increased testing also calls for a clearer definition of professional responsibilities. Professional use of tests could be limited, for example, to cases where it is clear that medical intervention or preventive measures will benefit the individual. At the very least, the fiduciary relationship involved in medical practice suggests that doctors should warn their patients not only about the availability of tests but also about the possible social consequences of being tested. Physicians may even be called upon to counsel patients not to obtain biological information when nothing can be done to ameliorate a suspected condition.

The power of testing also calls for clarifying the role of those medical professionals who work in nonclinical settings. Physicians in private companies or public schools currently operate under loosely constructed constraints. There are few ways to challenge their decisions. Their presumed loyalty to the ethical codes of their profession is belied by their conflicting obligations. As testing enhances their power, policies must be developed to verify professional judgments and to inform affected individuals so that their needs and interests will prevail. The professional responsibility of physicians should also extend to influencing institutional policy concerning the use of tests. They should educate administrators who use tests as to the appropriateness, statistical relevance, and practical limits of diagnostic information.

Considering the type of information revealed by new technologies and the development of data banks to store such information, it is critical to strengthen the laws and policies protecting privacy. Such policies must control access to information and limit its use so as to protect the rights of individuals. There are certainly precedents for such regulation. When Congress created the U.S. Census, it placed narrow restrictions on the use of the data collected by the Census Bureau. Similarly, the broad implications of biological data suggest that access to sensitive test results should be limited to those who have compelling social need for such information. And except in the most unusual circumstances, the individual tested should control the use of test results.

Finally, policies should be developed to prevent discrimination against those who are tested and found to be predisposed to disease. Those already employed or insured have some protection; but few policies or practices, such as high-risk insurance pools, subsidized health care services, or employment alternatives, are in place to protect those excluded on the basis of biological characteristics or to provide them with viable options.

Such social policies are imperative, for the most important implication of biological testing is the risk of expanding the number of people who simply do not fit. The growing scientific knowledge about genetics and the brain is serving to refine biological diagnosis, and tests have greatly improved in their sensitivity and accuracy. But even as tests increase in certainty and extend the range of what they can predict, questions of interpretation will remain. What degree of correlation between existing markers and subsequent physical or behavioral manifestations is necessary before taking social action such as exclusion from work, tracking in special education programs, or establishing competence to stand trial? How do we balance the institutional need for stability against the rights of the individual? What is to be defined as normal or abnormal, able or disabled, healthy or diseased? And whose yardstick should prevail?

Perhaps most important, even if there were perfect predictive information, we should not underestimate the dangers of a new eugenics. If biological tests are used to conform people to rigid institutional norms, we risk reducing social tolerance for the variation in human experience. We risk increasing the number of people defined as unemployable, uneducable, or uninsurable. We risk creating a biologic underclass.

GLOSSARY

ABA	American Bar Association
ACLU	American Civil Liberties Union
ACOG	American College of Obstetrics and Gynecology
AFP	Alpha-fetoprotein test
ALI	American Law Institute
AMA	American Medical Association
APA	American Psychiatric Association
BEAM	Brain Electrical Activity Mapping
CAT	Computer-Assisted Tomography
CEEG	Computer Electroencephalography
CMT	Charcot Marie Tooth Disease
CVS	Chorionic Villus Sampling
DRGs	Diagnostic Related Groups
DSM	Diagnostic and Statistical Manual of the American Psychiatric Association
EEOC	Equal Employment Opportunities Commission
ELISA	Enzyme–Linked Immunosorbent Assay (AIDS test)
ERISA	Employment Retirement Income Security Act
HLA	Human Leukocite Antigen
HMO	Health Maintenance Organization
MIB	Medical Information Bureau
MRI	Magnetic Resonance Imaging
NIOSH	National Institute for Occupational Safety and Health
OCAW	Oil, Chemical and Atomic Workers
OSHA	Occupational Safety and Health Administration
OTA	Office of Technology Assessment
PET	Positron Emission Tomography

PKU	Phenylketonuria
PMS	Premenstrual Syndrome
RFLP	Restriction Fragment Length Polymorphism
SPECT	Single Proton Emission Computed Tomography
SQUID	Superconducting Quantum Interference Devices

NOTES

Chapter 1 The New Diagnostics

1. Dana Fradon, *New Yorker*, June 29, 1987.

2. Alvan Feinstein, *Clinimetrics* (New Haven: Yale University Press, 1987).

3. U.S. Congress, Office of Technology Assessment, *Medical Testing and Health Insurance* (Washington, D.C.: GPO, 1988).

4. Charles R. Scriver, "Presidential Address," *American Journal of Human Genetics* 40 (1987): 199–211.

5. Orrie Friedman, quoted in Christopher Joyce, "Genes Reach the Medical Market," *New Scientist*, July 16, 1987: 48–51.

6. Dr. Robert Pasnau, "Presidential Address," *New Physician*, April 1988, p. 30.

7. National Institute of Mental Health, *Approaching the 21st Century: Opportunities for NIMH Neurosciences Research*, Report to Congress on the Decade of the Brain (Washington, D.C.: U.S. Department of Health and Human Services, January 1988).

8. Laura Souhrada, "Costly Equipment Moves into Psychiatric Care," *Hospitals* 62 (June 20, 1988): 76–77.

9. NIMH, *Approaching the 21st Century*.

10. Kathleen McAuliffe, "Predicting Diseases," *U.S. News and World Report*, May 25, 1987, pp. 64–69.

11. Orrie Friedman of Collaborative Research, quoted in *Hospitals*, February 20, 1986.

12. Michael Foucault, *Discipline and Punish* (New York: Vintage Books, 1979), pp. 183–84, 304.

13. Walter Reich, "Diagnostic Ethics: The Uses and Limits of Psychiatric Explanation," in *Ethical Issues in Epidemiologic Research*, Laurence Tancredi, ed. (New Brunswick, NJ: Rutgers University Press, 1986), pp. 37–69.

14. Joseph R. Gusfield, *The Culture of Public Problems* (Chicago: University of Chicago Press, 1981), suggests how scientific approaches reflect social and cultural perspectives about public problems. See also Jonathan Simon, "The Emergence of a Risk Society: Insurance Law and the State," *Socialist Review* 95 (September/October 1987): 63–89. He argues that social practices are increasingly shaped by an insurance model in which concerns about risk are met by aggregating individual experience and planning for risk in order to limit liability.

15. Alan Westin, *Information Technology in Democratic Society* (Cambridge: Harvard University Press, 1970).

16. *New York Times*, February 24, 1971.

17. For studies of the medicalization of social problems, see Peter Conrad and Rochelle Kern, eds., *The Sociology of Health and Illness: Critical Perspectives*, 2nd. ed. (New York: St. Martin's Press, 1986) and Peter Conrad and Joseph Schneider, eds., *Deviance and Medicalization* (St. Louis: C. V. Mosby, 1980).

18. Lenn and Madeleine Goodman, "Prevention—How Misuse of a Concept Undercuts Its Worth," *Hastings Center Report*, April 1986, pp. 26–38; and Deborah Stone, "The Resistible Rise of Preventive Medicine," *Journal of Health Politics, Policy and Law*, November 4, 1986, pp. 671–96.

19. Daniel J. Kevles, *In the Name of Eugenics* (New York: Alfred A. Knopf, 1985); Garland Allen, "The Misuse of Biological Hierarchies," *History and Philosophy of the Life Sciences*, May 2, 1983, pp. 105–28.

20. Stephen J. Gould, *The Mismeasure of Man* (New York: W. W. Norton, 1981); and Leon Kamin, *The Science and Politics of IQ* (New York: Lawrence Erlbaum, 1974).

21. Robert Proctor, *Racial Hygiene: Medicine Under the Nazis* (Cambridge: Harvard University Press, 1988).

22. Howard Kaye, *The Social Meaning of Biology* (New Haven, CT: Yale University Press, 1986).

23. Benno Muller-Hill, "Genetics After Auschwitz," *Holocaust and Genocide Studies* 2, no. 1 (1987): 3–20.

24. Linus Pauling, "Our Hope for the Future," in *Birth Defects*, M. Fishbein, ed. (Philadelphia: Lippincott, 1963).

25. Philip Handler, ed., *Biology and the Future of Man* (New York: Oxford University Press, 1970), pp. 916–27.

26. Bentley Glass, "Science: Endless Horizon or Golden Age?" *Science* 171 (1971): 23–29.

27. Leon R. Kass, "The New Biology: What Price Relieving Man's Estate?" *Science* 174 (November 19, 1971): 779–88.

28. See, for example, Daniel Koshland, Jr., Editorial, *Science* 235 (March 20, 1987): 1445.

29. See, for example, Arthur Jensen, *Genetics and Education* (New York: Harper & Row, 1972). Also Camille Benbow and Julian Stanley, "Sex Differences in Mathematical Reasoning: Fact or Artifact?" *Science* 210 (December 12, 1980): 1262–64.

30. Margery W. Shaw, "Conditional Prospective Rights of the Fetus," *Journal of Legal Medicine* 63 (1984): 63–116.

31. See discussion in Diane Paul, "Eugenic Origins of Clinical Genetics," mimeodraft, University of Massachusetts, Boston, 1988.

32. See discussion of this language in the President's Commission for the Study of Ethical Problems in Medicine #83-600502 (Washington, D.C.: GPO, 1983), and in Kaye, *The Social Meaning of Biology*.

33. E. O. Wilson, *Sociobiology* (Cambridge: Belknap Press, 1975).

34. *Time*, December 15, 1980; *Discover*, April 15, 1981; *Psychology Today*, February 1981; *U.S. News and World Report*, April 13, 1987, p. 59. For other examples, see Dorothy Nelkin, *Selling Science* (New York: W. H. Freeman, 1987).

35. James Q. Wilson and Richard J. Herrnstein, *Crime and Human Nature* (New York: Simon & Schuster, 1985).

36. Arthur Barsky, *Worried Sick: Our Troubled Quest for Wellness* (Boston: Little, Brown, 1988).

37. Nancy Shute, "How Healthy Is Your Family Tree?" *Hippocrates*, January/February 1988, pp. 88–89.

38. *New York Times*, August 16, 1988.

39. For analyses of the meaning of medical diagnoses, see Eliot Freidson, *The Profession of Medicine* (New York: Dodd, Mead, 1977) and Andrew Abbott, *The System of Professions* (Chicago: University of Chicago Press, 1988).

40. Some of the key studies in this social constructivist literature include Bruno Latour and Steve Woolgar's *Laboratory Life* (Beverly Hills, CA: Sage Publications, 1979); Karen Knorr-Cetina, *The Manufacture of Knowledge* (Oxford: Pergamon Press, 1981); and H. M. Collins, *Changing Order* (London: Sage Publications, 1985). Studies of the importance of social context in the development and use of scientific claims appear in Barry Barnes and David Edge, eds. *Science in Context* (Cambridge: MIT Press, 1982) and Bruno Latour, *Science in Action* (Cambridge: Harvard University Press, 1986).

Chapter 2 Defining Diagnosis

1. John H. Warner, *The Therapeutic Perspective: Medical Practice, Knowledge and Identity* (Cambridge: Harvard University Press, 1986), p. 58.

2. Stanley Reiser, *Medicine and the Reign of Technology* (Cambridge: Cambridge University Press, 1978).

3. American Medical Association, *Health Quackery* (pamphlet) (Chicago: AMA, 1968).

4. A. L. Cochrane, *Effectiveness and Efficiency: Random Reflections on Health Sciences* (London: Burgess and Sons, 1972).

5. Gordon H. Guyatt et al., "The Role of Before-After Studies of Therapeutic Impact in the Evaluation of Diagnostic Technologies," *Journal of Chronic Disabilities* 39, no. 4 (1986): 295–304.

6. Brian Jennett, "High Technology Medicine: How Defined and How Regarded," *Milbank Quarterly* 63, no. 1 (1985): 141-73.

7. Reiser, *Medicine and the Reign of Technology*.

8. Allan Brandt, *No Magic Bullet* (New York: Oxford University Press, 1985).

9. For a review of these techniques, see Sherman Elias and George Annas, *Reproductive Genetics and the Law* (Chicago: Yearbook Medicine Publishers, 1987).

10. Peter Rowley, "Genetic Screening: Marvel or Menace?" *Science* 225 (July 13, 1985): 139–44.

11. U.S. Congress, Office of Technology Assessment, *Mapping Our Genes* (Washington, D.C.: GPO, 1988). See also Ulf Landegren et al., "DNA Diagnostics—Molecular Techniques and Automation," *Science* 224 (October 14, 1988): 229–37.

12. Dr. Brenda Conner, quoted in *U.S. News and World Report*, May 25, 1987

13. N. D. Volkow and L. R. Tancredi, "Positron Emission Tomography: A Technology Assessment," *International Journal of Technology Assessment and Health Care* 2 (1986): 577–94.

14. E. R. John, L. S. Pricher, J. Fridman, and P. Easton, "Neurometrics: Computer Assisted Differential Diagnosis of Brain Dysfunctions," *Science* 239 (January 8, 1988): 162–69.

15. Herman Winick, "Synchrotron Radiation," *Scientific American* 257, no. 5 (November 1987): 88–99.

16. Marc Lappe, "The Limits of Genetic Inquiry," *Hastings Center Report*, August 1987, p. 8.

17. Genetic Technology News, "Market for DNA Probe Test for Genetic Diseases," *Genetic Technology News*, November 1986, pp. 6–7; see also U.S. Congress, Office of Technology Assessment, *Medical Testing* (Washington, D.C.: GPO, 1988), pp. 132–34.

18. Quoted in *U.S. News and World Report*, May 27, 1987.

19. Quoted in *Hospitals*, February 20, 1986.

20. *Collaborative Research Annual Reports*, 1986, 1987.

21. Quoted in *The Scientist*, June 27, 1988, p. 7.

22. Christopher Joyce, "Genes Reach the Medical Market," *New Scientist*, July 16, 1987.

23. John Tierney, "Thanks for Your Application, We'll Keep Your Fluids on File," *Hippocrates*, November/December 1987, pp. 33–34.

24. Leslie Roberts, "The Race for the Cystic Fibrosis Gene," *Science* 240 (8 April 1988): 141–45.

25. Michael McGinnis, cited in *U.S. News and World Report*, May 25, 1987.

26. Uriel Foa and David Margules, "The Opening of the Black Box: Is Psychology Prepared?" *Journal of Mind and Behavior* 4, no. 4 (Autumn 1983): 435–50.

27. David Cohen and Henry Cohen, "Biological Theories, Drug Treatments and Schizophrenia," *Journal of Mind and Behavior* 7, no. 1 (Winter 1986):11–36.

Chapter 3 Interpreting Tests

1. Dr. Mathon Trobe, a neurologist quoted in Robin Marantz Henig, "The Inner Landscape," *New York Times Magazine*, April 17, 1988, pp. 59–60.

2. Judith Swazey, "Phenylketonuria: A Case Study in Biomedical Legislation," *Journal of Urban Law* 48 (1971): 883–931.

3. J. A. Egeland and A. M. Hofstetter, "Amish Study I: Affective Disorder Among the Amish," 1976–1980. *American Journal of Psychiatry* 140 (1983): 36.

4. S. D. Detera-Wadleigh, W. H. Berettini, L. R. Goldin, D. Boorman, S. Anderson, and E. S. Gershon, "Close Linkage of c-HARVEY-RAS-1 and the Insulin Gene to Affective Disorder Is Ruled Out in Three North American Pedigrees," *Nature* 325 (February 1987): 806.

5. S. J. Reiser, *Medicine and the Reign of Technology* (Cambridge: Cambridge University Press, 1978), p. 189.

6. Ibid., pp. 191, 194.

7. L. Goldman et al., "The Value of the Autopsy in Three Medical Eras," *New England Journal of Medicine* 308 (1983): 1000–1005.

8. T. Kircher, J. Nelson, and H. Burdo, "The Autopsy as a Measure of Accuracy of the Death Certificate," *New England Journal of Medicine* 313 (1985): pp. 1263–69.

9. D. Rothschild and M. L. Sharp, "The Origin of Senile Psychosis: Neuropathic Factors and Factors of a More Personal Nature," *Diseases of the Nervous System* 2 (1941): 49–56.

10. L. N. Robbins, "Epidemiology: Reflections on Testing the Validity of Psychiatric Interviews," *Archive of General Psychiatry* 42 (1985): 918–24.

11. See A. Tversky and D. Kahneman, "Judgment Under Uncertainty: Heuristics and Biases," *Science* 185 (1974): 1124–36.

12. See review of such studies in David Faust and Jay Ziskin, "The Expert Witness in Psychology and Psychiatry," *Science* 241 (July 1, 1988): 31–35.

13. There is extensive statistical literature addressing such problems associated with Bayesian analysis. For example, see Bernard Rosner, *Fundamentals of Biostatistics*, 2nd ed. (Boston: Duxbury Press, 1986), pp. 54–58.

14. B. Tabakoff et al., "Differences in Platelet Enzyme Activity Between Alcoholics and Non-Alcoholics," *New England Journal of Medicine* 318 (1988): 134–39.

15. Theodore Reich, "Biological Marker Studies in Alcoholism," *New England Journal of Medicine* 318 (January 21, 1988): 181.

16. Ibid., p. 180.

Chapter 4 Diagnosis in the Health Care System

1. Society of Actuaries, *Record*, Annual Meeting, October 5–7, 1986, pp. 2943–62.

2. Paul Starr, *The Social Transformation of American Medicine* (New York: Basic Books, 1982); Bradford H. Gray, ed., *The New Health Care for Profit: Doctors and Hospitals in a Competitive Environment* (Washington, D.C.: Institute of Medicine, 1983).

3. U.S. Congress, Office of Technology Assessment, *AIDS and Health Insurance*, Staff Paper, February 1988.

4. Starr, *The Social Transformation of American Medicine*, pp. 393–98.

5. Daniel Waldo, Katharine R. Levit, and Helen Lazenby, "National Health Expenditures, 1985," *Health Care Financing Review* 8, no. 1 (Fall 1986): 4.

6. See Mark Holoweiko, "What the Pre-Paid Care 'Boom' Looks Like from the Inside," *Medical Economics* 21 (January 1985): 103–14.

7. Dan Ermann and Jon Gabel, "Multihospital Systems: Issues and Empirical Findings," *Health Affairs* 53 (Spring 1984): 50–61.

8. Starr, *The Social Transformation of American Medicine*, p. 431, and *Wall Street Journal*, October 9, 1977.

9. Starr, *The Social Transformation of American Medicine*, p. 432.

10. See Jeffrey A. Wasserman, "How DRGs Work," *Hastings Center Report*, October 1983, pp. 23–25.

11. Phil Brown, "Diagnostic Conflict and Contradiction in Psychiatry," *Journal of Health and Social Behavior* 28 (March 1987): 43.

12. Steven A. Schroeder, "Strategies for Reducing Medical Costs by Changing Physicians' Behavior," *International Journal of Technology Assessment in Health Care* 3 (1987): 39–50.

13. Sherman Elias and George Annas, "Routine Pre-Natal Genetic Screening," New England Journal of Medicine 317, no. 22 (November 26, 1987): 1407–8.

14. *Curlender v. BioScience Laboratories* 106 Cal. App.3d 811, 165 Cal. Rptr. 477 (1980).

15. *Schroeder v. Perkel*, 87 N.J. 53, 432 A.2d 834 (1981).

16. Edward P. Richards and Katherine C. Rathbun, *Medical Risk Management: Preventive Legal Strategies for Health Care Providers* (Rockville, MD: Aspen Publications, 1983), p. 27.

17. AMA Special Task Force on Professional Liability and Insurance, *Professional Liability in the 1980s* (Chicago: AMA Report 1:16, October 1984).

18. See *New York Times*, July 28, 1988.

19. I. Weinstein, "The Future of the M.D. in a Price-Competitive, Price-Driven Market," *Topics in Health Care Financing* 11 (1984): 91.

20. Roz Lasker, "The Demise of a Private Practice," *American College of Physicians, Observer* 7 (March 1987): 20.

21. U.S. Congress, Office of Technology Assessment, *Medical Testing and Health Insurance* (Washington, D.C.: GPO, 1988), pp. 3–4.

22. Neil A. Holtzman, "Recombinant DNA Technology, Genetic Tests and Public Policy," *American Journal of Human Genetics* 42 (April 1988): table 4, pp. 624–32.

23. Mark Slesinger et al., "The Privatization of Health Care and Physicians' Perceptions of Access to Hospital Services," *Milbank Quarterly* 65 (1987): 25–58.

24. See Sheldon Greenfield and M. Colin Jordan, "The Clinical Investigation of Lymphadenopathy in Primary Care Practice," *Journal of the American Medical Association* 240, no. 13 (1978): 1388–91.

25. Stephen G. Pauker and Jerome P. Kassirer, "Decision Analysis," *New England Journal of Medicine* 316, no. 5 (1987): 250–58.

26. E. J. M. Campbell, "The Diagnosing Mind," *Lancet*, April 11, 1987, pp. 849–51.

27. L. R. Tancredi and M. Edlund, "Are Conflicts of Interests Endemic to Psychiatric Consultation?" *International Journal of Law and Psychiatry* 6 (1983): 298–316.

28. Dorothy Nelkin, *Selling Science: How the Press Covers Science and Technology* (New York: W. H. Freeman, 1987), pp. 132–69.

29. *Bergen Record*, July 28, 1988.

30. B. Dickens, "Abortion, Amniocentesis and the Law," *American Journal of Comparative Law* 34 (1986): 249.

31. A. Milunsky, "Genetic Counseling: Prelude to Prenatal Diagnosis," in *Genetic Diseases and the Fetus*, A. Milunsky, ed. (New York: Plenum Press, 1986), pp. 2–24.

32. Barbara Katz Rothman, *The Tentative Pregnancy* (New York: Viking Press, 1986).

33. Survey by John Fletcher, reported in *Medical World News*, September 28, 1987.

34. *New York Times*, July 18, 1988.

35. Daniel J. Kevles, *In the Name of Eugenics* (New York: Alfred A. Knopf, 1985), p. 277.

36. *In re Moore* 28 NC 95 (1976).

37. Kevles, *In the Name of Eugenics*, p. 277.

38. Letters from E. B. Hook, M.D., from New York State Department of Health to New York State Laboratories, April 2, 1981; and from Alice Starr, New York State Department of Health, July 29, 1987. They were given to the authors by Dr. Sandra Wolman of the New York University Medical School.

39. Joan K. Burns et al., "Impact of PKU on the Reproductive Patterns in Collaborative Study Families," *American Journal of Human Genetics* 19 (1984): 515–24.

40. *Boston Globe*, July 29, 1985.

41. G. L. Klerman, "The Significance of the DSM III in American Psychiatry," in *International Perspectives on DSM III*, R. L. Spitzer, J. Williams, and A. E. Skodal, eds. (Washington, D.C.: American Psychiatric Press, 1983), pp. 3–25; and N. C. Andreasen, *The Broken Brain, The Biological Revolution in Psychiatry* (Cambridge: Harvard University Press, 1984).

42. Laura Souhrada, "Costly Equipment Moves into Psychiatric Care," *Hospitals* 62 (June 20, 1988): 76.

43. Jack H. Blaine, "AIDS: Regulatory Issues for Life and Health Insurers," *AIDS and Public Policy Journal* 2 (Winter 1987): 2–10.

44. Op-Ed, *New York Times*, September 19, 1987.

45. U.S. Congress, *Medical Testing and Health Insurance*.

46. Blaine, "AIDS," p. 3.

47. Daniel Case, "AIDS: How Many Icebergs?" *The Actuary* (newsletter of the Society of Actuaries) 21, no. 2 (February 1987): 1, 5.

48. Cited in *Boston Globe*, March 7, 1988.

49. James Gusella et al., "A Polymorphic DNA Marker Genetically Linked to Huntington's Disease," *Nature* 306 (November 7, 1983): 234–38.

50. Colin B. Begg, Robert A. Greens, and Boris Iglewicz, "The Influence of Uninterpretability on the Assessment of Diagnostic Tests," *Journal of Health and Social Behavior* 28 (March 1987): 37–50.

51. See Deborah Stone, "The Resistible Rise of Preventive Medicine," *Journal of Health Politics* 11, no. 4 (1986): 671–96.

Chapter 5 Testing in the Workplace: Predicting Performance and Health

1. Anthony Mazzochi, "Working for Your Life," lecture at Oil, Chemical and Atomic Workers Trade Union Conference, St. Louis, September 1980.

2. Statistical information on occupational health in these industries is available in reference reports published by the Bureau of Labor Statistics and the U.S. Department of Labor.

3. *Chemical Week*, January 13, 1982, p. 40.

4. Occupational Safety and Health Act, 1970.

5. Sheila Jasanoff, "The Misrule of Law at OSHA," in *The Language of Risk*, D. Nelkin, ed. (Beverly Hills: Sage Publications, 1985).

6. Cited in Richard Severo, "Federal Mandate for Gene Tests Disturbs U.S. Job Safety Official," *New York Times*, February 6, 1980.

7. See Peter Barth and H. A. Hunt, *Workers Compensation and Work-Related Illnesses and Diseases* (Cambridge: MIT Press, 1980).

8. John D. Blum, "Corporate Liability from In-house Medical Malpractice," *St. Louis University Law Review* 22, no. 3 (1978): 433–51.

9. U.S. Chamber of Commerce, *Employee Benefits—1986* (Washington, D.C.: GPO, 1988).

10. Cited in Milton Freudenheim, "Business and Health," *New York Times*, December 1, 1987.

11. Laird L. Miller, "Honeywell, Inc.," *Health Care Financing Review*, 1986 Annual Supplement, p. 96.

12. Jane Sasseen, "The Bad Seed," *Forbes* (April 11, 1983): 160–62.

13. Occupational Safety and Health Administration, *Occupational Safety and Health Reporter*, May 7, 1981, p. 1526.

14. Wendy Williams, "Firing Women to Protect the Fetus," *Georgetown Law Journal* 69 (1981): 641–704.

15. *Christman v. American Cyanamid Co.*, #30–0024, Northern District of West Virginia filed 1.30.80, settled.

16. *Oil, Chemical and Atomic Workers International Union v. American Cyanamid*, 741 F2d 444 (1984).

17. Ibid.

18. Elaine Draper, "High Risk Workers or High Risk Work," *International Journal of Sociology and Social Policy* 6, no. 4 (1986): 12–28. Also see Dorothy Nelkin and Michael S. Brown, *Workers at Risk* (Chicago: University of Chicago Press, 1984).

19. Thomas McCarity and Elinor Schroeder, "Risk-Oriented Employment Screening," *Texas Law Review* 59, no. 6 (August 1981): 1000–1076.

20. William E. Powles and W. Donald Ross, "Industrial and Occupational Psychiatry," in *American Handbook of Psychiatry*, vol. 3, Silvano Arieti, ed. (New York: Basic Books, 1966), pp. 595–96.

21. DuPont Employee Relations Department, *Occupational Medicine Program* (Wilmington, DE: DuPont, 1981). For a general review of corporate medicine, see Dorothy Nelkin, "Ethical Conflicts in Corporate Medicine," in *The Language of Risk*, Nelkin, and Diane Chapman Walsh, eds. *Corporate Physicians: Between Medical Care and Management* (New Haven: Yale University Press, 1987).

22. Mark Rothstein, *Medical Screening of Workers* (Washington, D.C.: Bureau of National Affairs, 1984), p. 10.

23. Vivienne Walters, "Company Doctors' Perception of and Responses to Conflicting Pressures from Labor and Management," *Social Problems* 30, no. 1 (October 1982): 1–12.

24. *Newsweek*, May 5, 1986, p. 46.

25. Deborah A. Stone, "The Resistible Rise of Preventive Medicine," *Journal of Health Politics, Policy and Law* 11, no. 4 (1986): 671–96. Stone sees the tendency to screen workers as part of the general ideology of preventive medicine.

26. Mark Rothstein, "Employee Selection Based on Susceptibility to Occupational Illness," *Michigan Law Review* 81 (May 1983): 1379–96.

27. Sally Guttmacher, "The Ethics of Workplace Screening," *Business and Health*, March 1984, pp. 23–26.

28. Stanley Reiser, "The Emergence of the Concept of Screening for Disease," *Milbank Memorial Fund Quarterly* 56, no. 4 (1978): 78.

29. HEW National Occupational Hazard Survey (Washington, D.C.: GPO, 1977), pp. 77–78. See also Mark A. Rothstein, "Medical Screening: A Tool with Broadening Use," *Business and Health* (October 1986): 8.

30. M. Laurens Rowe, "Are Routine Spine Films on Workers in Industry Cost- or Risk-Benefit Effective?" *Journal of Occupational Medicine* 24 (January 1982): 41–43.

31. Richard C. Sweetland and Daniel Keyser, eds., *Tests: A Comprehensive Reference for Assessments in Psychology, Education and Business* (Kansas City: Test Corporation of America, 1986).

32. Marc Reisch, "Major Chemical Producers Toughen Stance on Drug Abuse," *Chemical and Engineering News*, August 3, 1987, pp. 7–9.

33. *Genetic Engineering News*, March 1987, pp. 14, 34.

34. *Wall Street Journal*, NBC news poll, June 18, 1987.

35. Joyce C. Hogan and Edward Bernacki, "Developing Job-Related Preplacement Medical Examinations," *Journal of Occupational Medicine* 23 (July 1981): 474.

36. Leo Uzych, "Genetic Testing and Exclusionary Practices in the Workplace," *Journal of Public Health Policy* 7, no. 1 (Spring 1986): 37–57.

37. *Chemical and Engineering News*, February 18, 1980, pp. 35–37.

38. Joan E. Bertin, "Discrimination Against Women of Childbearing Capacity," address at Hastings Center, 1982.

39. Quoted in Ronald Bayer, "Women, Work and Reproductive Hazards," *Hastings Center Report*, October 1982, p. 17.

40. *Chemical and Engineering News*, February 11, 1980, p. 30.

41. U.S. Congress, Office of Technology Assessment, *The Role of Genetic Testing in the Prevention of Occupational Disease* (Washington, D.C.: OTA, 1983).

42. *Sterling Transit Co. v. Fair Employment Practice Commission*, 175 Cal. Rptr. 548, 121 Cal. App.3d 791 (1981). For a review of legal issues raised by genetic testing, see OTA, *The Role of Genetic Testing.*

43. *E. E. Black, LTD v. U.S. Department of Labor et al.* 497 F.Supp. 1088 (1980).

44. Edward Calabrese, "Ecogenetics: The Historical Foundation and Current Status," *Journal of Occupational Medicine* 28, no. 10 (October 1986).

45. National Academy of Sciences, *Principles of Evaluating Chemicals in the Environment* (Washington, D.C.: GPO, 1975)

46. For a review of the arguments concerning genetic screening and monitoring, see Elaine Draper, "High Risk Workers or High Risk Work," *International Journal of Sociology and Social Policy* (Fall 1986): 12–28.

47. Jennifer Ratcliffe et al., "The Prevalence of Screening in Industry: Report from the National Institute for Occupational Safety and Health National Occupational Survey," *Journal of Occupational Medicine* 28, no. 10 (October 1986): 911.

48. OTA, *The Role of Genetic Testing.*

49. Ibid.

50. The series, by Richard Severo, appeared in the *New York Times* on February 3, 4, and 6, 1980; U.S. Congress, House of Representatives Subcommittee on Investigations and Oversight of the Committee on Science and Technology, *Genetic Screening and the Handling of High Risk Groups in the Workplace*, 99th Cong., 1st sess., October 14–15, 1981; and 2nd sess., October 6, 1982.

51. Morton Hunt, "The Total Gene Screen," *New York Times Magazine*, January 19, 1986, pp. 33ff.

52. Draper, "High Risk Workers."

53. OTA, *The Role of Genetic Testing.*

54. See Edith Canter, "Employment Discrimination: Implications of Genetic Screening under Title VII and the Rehabilitation Act," *American Journal of Law and Medicine* 10, no. 3 (1984): 323–47.

55. For reviews of these disputes, see Jess McKenzie, "Evaluation of the Hazards of Sickle Trait in Aviation," *Aviation, Space and Environmental Medicine*, August 1977, pp. 753–62; Anne Hoiberg, John Ernst, and David Uddin, "Sickle Cell Trait and Glucose-6-Phosphate Dehydrogenase Deficiency," *Archives of Internal Medicine* 141 (October 1981): 1485–88; C. Holden, "Air Force Challenge on Sickle Trait Policy," *Science* 211 (January 16, 1986): 257; and "Newsletter," *Military Medicine* 152 (November 1987): A6.

56. Testimony of Bruce W. Karrh, M.D., Corporate Medical Director of DuPont, U.S. Congress, *Genetic Screening*, October 15, 1981.

57. U.S. Congress, *Genetic Screening*.

58. Focus Technologies, *Annual Reports*.

59. *Industry Week*, June 1, 1987; p. 44.

60. Ibid.

61. *Child Protection Group v. Cline*, 17296 (W. Va. Sup. Ct., November 12, 1986).

62. Quoted in *The Wall Street Journal*, February 24, 1986.

63. Elizabeth Whelan, "Chemicals and Cancerphobia," *Society*, March/April 1981, p. 7.

64. Bernard D. Davis, statement at conference on Genetic Influences on Responses to the Environment, National Academy of Sciences, Institute of Medicine, Washington, D.C., 1980, pp. 232–33.

65. Peter Conrad, "Wellness in the Workplace," *Milbank Quarterly* 65, no. 2 (1987): 269–72; Draper, "High Risk Workers."

Chapter 6 The Schools: Testing for Learning Problems

1. For a description of these debates see Michael Katz, *The Irony of Early School Reform* (Cambridge: Harvard University Press, 1968); quotes are from pp. 181 and 215.

2. Arthur Gates, "The Role of Personality Maladjustment in Reading Disability," in *Children with Reading Problems*, Gladys Natchez, ed. (New York: Basic Books, 1968).

3. See, for example, Paolo Freire, *Pedagogy of the Oppressed* (New York: Herder and Herder, 1971); and Annette Rubinstein, ed., *Schools Against Children* (New York: Monthly Review Press, 1970).

4. Irving Harris, *Emotional Blocks to Learning: A Study of the Reasons for Failure in School* (Glencoe, Ill.: The Free Press, 1962).

5. Randi Londer, "Cracking the Code," *Child* (January/February 1988): 44–46, 102–3.

6. Cecile Reynolds, "Critical Measurement Issues in Learning Disabilities," *Journal of Special Education* 18 (1985): 451–75.

7. For a historical discussion of this tendency, see Katz, *The Irony of Early School Reform*.

8. See discussion in Dorothy Nelkin, *Selling Science* (New York: W. H. Freeman, 1987), pp. 21–24.

9. National Commission on Excellence in Education, *A Nation at Risk* (Washington, D.C.: GPO, 1983), p. 5. Also see Ira Shor, *Culture Wars: School and Society in the Conservatory Restoration, 1979–1984* (London: Routledge & Kegan Paul, 1976).

10. Hugh Mehan, Alma Hertweck, and J. Lee Meihls, *Handicapping the Handicapped: Decision Making in Students' Educational Careers* (Stanford, CA: Stanford University Press, 1986).

11. See Stephen Jay Gould, *The Mismeasure of Man* (New York: W. W. Norton, 1981); Daniel J. Kevles, *In the Name of Eugenics* (Berkeley and Los Angeles: University of California Press, 1985); and Nancy Stepan, *The Idea of Race in Science* (London: Macmillan Press, 1982).

12. Gould, *The Mismeasure of Man*, pp. 155, 179.

13. Kevles, *In the Name of Eugenics*, p. 82.

14. See N. J. Block and G. Dworkin, *The IQ Controversy* (New York: Pantheon, 1976); H. J. Eysenck, *The IQ Argument: Race, Intelligence and Education* (New York: Library Press, 1971); A. R. Jensen, "How Much Can We Boost IQ and Scholastic Achievement?" *Harvard Educational Review* 33 (1969): 1–23; and L. J. Kamin, *The Science and Politics of IQ* (New York: Halstead Press, 1974).

15. Cecile Reynolds and Terry Gutkin, *The Handbook of School Psychology* (New York: Wiley, 1982), p. 212.

16. Arthur Jensen, "Objectivity and the Genetics of IQ," *Phi Delta Kappan* 66, no. 4 (1984): 284–86.

17. From C. E. Gill, R. Jardine, and N. G. Martin, "Further Evidence for Genetic Influences on Educational Achievement," *British Journal of Educational Psychology* 55 (1985): 240–50. Also see K. Richardson, "Genotype-Phenotype Relations in Models of Educational Achievement: A Response to Gill et al.," *British Journal of Educational Psychology* 57 (1987): 1–8, which questions the mathematical model of genotype-phenotype interaction used by Gill et al. H. J. Eysenck, "A New View of Human Intelligence," *British Journal of Educational Psychology* 56 (February 1986): 106–8.

18. Peter Schrag and Diane Divoky, *The Myth of the Hyperactive Child* (New York: Dell, 1975), pp. 171, 16.

19. D. D. Hammil, J. Leigh, G. McNutt, and S. C. Larsen, "A New Definition of Learning Disabilities," *Learning Disability Quarterly* 4 (1981): 336.

20. This case was reconstructed from *Derry News*, May 4, 1988; *Union Leader* (Manchester, NH), April 27, May 25, and June 8, 1988; *Boston Globe*, April 29, May 25, and June 6, 1981; and author's interview with Michael Jessop, Casey's father.

21. Diane McGuinness, *When Children Don't Learn: Understanding the Biology and Psychology of Learning Disabilities* (New York: Basic Books, 1985), chaps. 9 and 10, pp. 168–230.

22. Reynolds and Gutkin, *The Handbook of Social Psychology*, p. 1025.

23. Debra Viadero, "Debate Grows on Classroom's Magic Pill," *Education Week* 7, No. 7 (October 21, 1987): 1–2.

24. Reynolds and Gutkin, *The Handbook of Social Psychology*, pp. 1025, 1026.

25. Schrag and Divoky, *The Myth of the Hyperactive Child*, p. 6.

26. See R. W. Peterson, "Great Expectations: Collaboration Between the Brain Sciences and Education," *American Biology Teacher* 46, no. 2 (February 1984): 74–80; also Jeanne S. Chall and Rita W. Peterson, "The Influence of Neuroscience upon Educational Practice," in *The Brain Cognition and Education*, Sarah Friedman, Kenneth A.

Klivington, and Rita Peterson, eds. (California: Academic Press, 1986), pp. 287–318; L. Hart, *Human Brain and Human Learning* (New York: Longman, 1983); and H. T. Epstein, "Developmental Biology and Disorders of Reading," in *Reading Disorders*, R. N. Malatesha and P. G. Aaron, eds. (New York: Academic Press, 1982); S. A. Mednick and J. Volavka, "Biology and Crime," in *Crime and Justice, An Annual Review of Research*, vol. 2, N. Morris and M. Tonry, eds. (Chicago: University of Chicago Press, 1980).

27. Jerome Kagan, J. S. Reznick, and N. Shidman, "Biological Bases of Childhood Shyness," *Science* 240 (April 8, 1988): 167–71.

28. *Harvard Magazine*, January/February 1988, p. 6.

29. *Boston Globe*, April 19, 1988.

30. For a review of this literature, see Sadie Decker and Bruce Bender, "Converging Evidence for Multiple Genetic Forms of Reading Disability," *Brain and Language* 33 (1988): 197–215.

31. Bruce Q. Bender, "Cognitive Development of Children with Sex Chromosome Abnormalities," in *Genetics and Learning Disabilities*, Shelly D. Smith, ed. (San Diego: College Hill Press, 1986).

32. Friedman, Klivington, and Peterson, *The Brain Cognition and Education*, p. 292.

33. Deborah M. Barnes, "Fragile X Syndrome and Its Puzzling Genetics," *Science* 243 (January 17, 1989): 171–72.

34. National Institute of Mental Health, *Approaching the 21st Century: Opportunities for NIMH Neuroscience Research*, Report to Congress on the Decade of the Brain (Washington, D.C.: U.S. Department of Health and Human Services, January 1988).

35. Friedman, Klivington, and Peterson, *The Brain Cognition and Education*, p. 300.

36. E. R. John, L. S. Prichep, J. Fridman, and P. Easton, "Neurometrics: Computer-Assisted Differential Diagnosis of Brain Dysfunctions," *Science* 239 (January 8, 1988): 162–69.

37. National Institute of Mental Health, *Approaching the 21st Century*.

38. Ibid.

39. Frank R. Vellutino, "Dyslexia," *Scientific American* 256 (March 1987): 34–41.

40. *New York Times*, August 18, 1987.

41. See, for example, *New York Times*, November 11, 1984.

42. Authors' interview with Dr. Bernard Feierstein, school psychologist, Ithaca, N.Y., school district, July 22, 1987.

43. David Rosner, "Health Care for the 'Truly Needy': 19th Century Origins of the Concept," *Milbank Quarterly* 60 (Summer 1982): 355–85.

44. See listing and discussion in Reynolds and Gutkin, *The Handbook of Social Psychology*, pp. 1023–39.

45. George Bear and Preston D. Modlin, "Gesell's Developmental Testing: What Purpose Does It Serve?" *Psychology in the Schools* 24 (January 1987): 40–44.

46. Peggy L. Anderson, Mary E. Cronin, and James H. Miller, "Referral Reasons for Learning-Disabled Students," *Psychology in the Schools* 23 (October 1986): 388.

47. See Kenneth A. Kavale and Steven R. Forness, "The Far Side of Heterogeneity: A Critical Analysis of Empirical Subtyping Research in Learning Disabilities," *Journal of Learning Disabilities* 20, no. 6 (June/July 1987): 374–82. Referral statements reviewed by Anderson, Cronin, and Miller, "Referral Reasons," emphasized disruptive classroom behavior in 58 percent of student referrals. On tolerance for learning difficulties, see B.

Algozzine, S. Christenson, and J. Ysseldyke, "Probabilities Associated with the Referral to Placement Process," *TEASE* 5, no. 3 (1982): 19–23.

48. Gerald S. Coles, *The Learning Mystique* (New York: Pantheon Books, 1988), p. 14.

Chapter 7 The Medical Empowering of Legal Institutions

1. *Barefoot v. Estelle*, 103 S.Ct. 3383 (1983). A previous Supreme Court case had also called into question the use of psychiatric evidence. In this case, however (*Estelle v. Smith*, 451 U.S. 454 [1981]), the court made no decision on the substantive merit of psychiatric information. The psychiatric tests were not admitted on procedural grounds (the defendant had not been appropriately informed that psychiatric evaluation would be used at the sentencing phase).

2. D. Faust and J. Ziskin, "The Expert Witness in Psychology and Psychiatry," *Science* 241 (July 1, 1988): 31–35.

3. J. L. Peterson, J. P. Ryan, P. J. Houlder, and S. Mihajlovic, "The Uses and Effects of Forensic Science in the Adjudication of Felony Cases," *Journal of Forensic Sciences* 3, no. 6 (November 1987): 1730–53.

4. A number of sources discuss the tests for insanity. See "The Insanity Defense: ABA and APA Proposals for Change," *Medical Disability Health Reporter* 7 (1983): 136–211, and D. Wexler, "Redefining the Insanity Problem," *George Washington Law Review* 53 (1985): 528–61.

5. P. Low, J. Jeffries, and R. Bonnie, *The Trial of John W. Hinckley* (New York: Foundation Press, 1986), p. 119; and M. Moore, "Causation and Its Excuses," *California Law Review* 73 (1985): 1091–1149.

6. L. Tancredi, J. Lieb, and A. Slaby, *Legal Issues in Psychiatric Care* (New York: Harper & Row, 1975).

7. M'Naghten Case (1843), House of Lords, 8 Eng. Rep. 718 (H6).

8. *Durham v. U.S.*, 214 F.2d 862 (D.C. Cir., 1954).

9. *Carter v. U.S.*, 252 F.2d 862 (D.C. Cir., 1957).

10. *McDonald v. U.S.*, 312 F.2d 847 (D.C. Cir., 1962).

11. *U.S. v. Brawner*, 471 F.2d 969 (1972).

12. ALI Model Penal Code (1955), sec. 4.01C1 (tentative draft #4).

13. *Washington v. U.S.*, 390 F.2d 444 (1967).

14. Low, Jeffries, and Bonnie, *John W. Hinckley*, p. 117; Wexler, "Redefining the Insanity Problem," p. 529.

15. G. Miller, "Prohibiting Diagnosis in Insanity Trials," *Psychiatry* 49 (1986): 131–43.

16. Low, Jeffries, and Bonnie, *John W. Hinckley*, pp. 27–30; Miller, "Prohibiting Diagnosis," p. 134.

17. C. Jones, "Men of Science v. Men of Law: Some Comments on Recent Cases," *Medicine, Science and Law* 26 (1986): 13–16.

18. Low, Jeffries, and Bonnie, *John W. Hinckley*, p. 119.

19. Sander Gilman, *Disease and Representation* (Ithaca, NY: Cornell University Press, 1988), p. 15.

20. D. Slater and V. Hans, "Public Opinion of Forensic Psychiatry Following the Hinckley Verdict," *American Journal of Psychiatry* 141 (1984): 675–79.

21. Phillip Resnick, "Perceptions of Psychiatric Testimony: A Historical Perspective on the Hysterical Invective," *Bulletin of the American Academy of Psychiatry and the Law* 14 (1986): 203–19.

22. *Harris Survey: Public Confidence in Institutions Remains Low*, released November 13, 1972; for a historical examination of criminal insanity and the role of psychiatry, see Roger Smith, "Criminal Insanity: From a Historical Point of View," *Bulletin of the American Academy of Psychiatry and the Law* 11 (1983): 27–34.

23. Resnick, "Perceptions of Psychiatric Testimony."

24. P. Talmadge, "Toward a Reduction of Washington Appellate Court Case Loads and More Effective Use of Appellate Court Resources," *Gonzaga Law Review* 21 (1985/86): 21–46.

25. Ibid.

26. M. Luskin and R. Luskin, "Why So Fast, Why So Slow?: Explaining Case Processing Time," *Journal of Criminal Law and Criminology* 77 (1986): 190–214.

27. M. Moore, *Law and Psychiatry: Rethinking the Relationship* (Cambridge: Cambridge University Press, 1984), p. 100.

28. Gilman, *Disease and Representation*.

29. *In Matter re: Torsney v. Gold*, 74 N.Y.2d 667, 394 N.E.2d 262 (1972).

30. *People v. Wright*, 648 P.2d 665 (Colo. 1982).

31. *Thompson v. Crawford*, 479 So.2d 169 (Fla. 3rd Dist. App. 1985).

32. *Nacher v. State*, 465 So.2d 598 (Fla. 3rd Dist. App. 1985).

33. Comment, "Resurrection of the Ultimate Issue Role: Federal Rule of Evidence 704(b) and the Insanity Defense," *Cornell Law Review* 72 (1987): 620–40.

34. "The Insanity Defense," p. 138.

35. Ibid., p. 145.

36. *Addington v. Texas*, 99 S.Ct. 1804, 1811 (1979).

37. *Ake v. Oklahoma*, 470 U.S. 68, 81 (1985).

38. "The Insanity Defense," p. 146.

39. The Insanity Defense Reform Act of 1984, 18 U.S.C. § 20 (1984); "The Insanity Defense."

40. Federal Rule of Evidence 704(B).

41. Jones, "Men of Science," p. 13.

42. *Frye v. United States*, 293 F. 1013 (D.C. Cir. 1923).

43. C. Boorse, "Premenstrual Syndrome and Criminal Responsibility," in *Premenstrual Syndrome*, B. Ginsberg and B. Carter, eds. (New York: Plenum Press, 1987), p. 81.

44. T. Riley, "Premenstrual Syndrome As a Legal Defense," *Hamline Law Review* 9 (1986): 193–202.

45. J. Trimble and M. Fay, "PMS in Today's Society," *Hamline Law Review* 9 (1986): 183–91.

46. S. Bird, "Neuroscience Research and Premenstrual Syndrome: Scientific and Ethical Concerns," in *Premenstrual Syndrome*, p. 34.

47. R. DiLiberto, "Premenstrual Stress Syndrome Defense: Legal, Medical, Social Aspects," *Medical Trial Quarterly* 33 (1987): 351–59.

48. Riley, "Premenstrual Syndrome As Legal Defense," p. 201.

49. K. Heggestad, "The Devil Made Me Do It: The Case Against Using Premenstrual Syndrome As a Defense in a Court of Law," *Hamline Law Review* 9 (1986): 155–82.

50. DiLiberto, "Premenstrual Stress Syndrome Defense," p. 358.

51. *Tice v. Richardson* 7 Kan. App. 2d 2507, 644 P.2d 490 (1982); 37 A.L.R. 4th 167.

52. Ibid., p. 490.

53. *State v. Blackman*, 662 P.2d 1183 (Kan. 1983).

54. D. Kaye and R. Kanwischer, "Admissibility of Genetic Testing in Paternity Litigations: A Survey of State Statutes," *Family Law Quarterly* 22, no. 2 (1988): 109–15. Eight of forty-seven statutes find presumption of paternity when probability of paternity is 95 to 99 percent.

55. D. Raskin, "Polygraph in 1986: Scientific, Professional, and Legal Issues Surrounding Application and Acceptance of Polygraph Evidence," *Utah Law Review* 1986 (1986): 29–74.

56. W. Curren and E. D. Shapiro, *Law, Medicine and Forensic Science* (Boston: Little, Brown, 1982), p. 165.

57. *United States v. Ridling*, 350 F. Supp. 90 (Mich. 1972); and Raskin, "Polygraph in 1986."

58. B. Dodd, "DNA Fingerprinting in Matters of Family and Crime," *Medicine, Science and Law* 26 (1986): 6–7.

59. *Tommie Lee Andrews v. State of Florida*, No. 87–2166, slip opinion (Fla. App., 5 Dist., Oct. 20, 1988).

60. *New York Times*, October 20, 1988.

61. *State v. Carlson*, 267 N.W. 2d 170 (Minn. 1978).

62. J. L. Marx, "DNA Fingerprinting Takes the Witness Stand," *Science* 240 (June 17, 1988): 1616–18.

63. U. K. Johnson, "Immigration Authorities May Use DNA Fingerprinting," *Nature* 39 (1987): 5.

64. *Boston Globe*, June 17, 1988, p. 14; *Hastings Center Report* 18 (October/November 1988): 3–4.

65. Patricia Jacobs et al., "Aggressive Behavior, Mental Subnormality and the XYY Male," *Nature* 208 (1965): 1351–52.

66. Comment, "The XYY Chromosomal Abnormality: Use and Misuse in the Legal Process," *Harvard Journal on Legislation* 9 (1972): 469–97.

67. Ibid., p. 481.

68. *Kaimowitz v. Dept. of Mental Health*, 370 U.S. 660 (1962).

69. N. Volkow and L. Tancredi, "Neural Substrates of Violent Behavior: A Preliminary Study with Positron Emission Tomography," *British Journal of Psychiatry* 151 (1987): 668–73. See also B. White, "Biological Causes for Violent Behavior: Research Could Affect Legal Decisions," *Texas Bar Journal* 50 (1987): 446.

70. Volkow and Tancredi, "Neural Substrates."

71. Discussion at the International Congress of Law and Mental Health, Montreal, June 17, 1988.

72. C. R. Jeffery in collaboration with R. V. Del Carmen and J. D. White, *Attacks on the Insanity Defense: Biological Psychiatry and New Perspectives on Criminal Behavior* (Springfield, Ill.: Charles C Thomas, 1985), p. 82.

73. *People v. Barry Wayne McNamara*, Santa Barbara County Superior Court, December 24, 1985 (unpublished case). Also, correspondence with Michael McGrath, Public Defender, Santa Barbara, California.

74. Personal communication, Dr. Monte S. Buchsbaum, University of California at Irvine, August 1988; and Mr. Robert Erwin, the attorney representing Mr. Hodge in the malpractice action, September 1988.

75. J. Stewart, "Billy Clubbed: Brain-Damaged Man Wins $2.7-Million Award," *Los Angeles Times*, pt. 2, December 6, 1987, pp. 1, 10, 11.

76. Jeffery et al., *Attacks on the Insanity Defense*, p. 117.

77. M. P. I. Weller, "Medical Concepts in Psychotherapy and Violence," *Medicine, Science and Law* 26 (1986): 131–43.

Chapter 8 Social Control Through Biological Tests

1. U.S. Congress, Office of Technology Assessment, *Medical Testing and Health Insurance*, OTA–H–384 (Washington, D.C.: GPO, 1988), p. 103.

2. Reviews of the sociological literature on social control can be found in Jack P. Gibbs, ed., *Social Control: Views from the Social Sciences* (Beverly Hills, CA: Sage Publications, 1982); medicine as a source of social control is analyzed in Peter Conrad, *Deviance and Medicalization* (St. Louis: C. V. Mosby, 1980), Irvin V. Zola, "Medicine As an Institution of Social Control," *Social Review* 20 (November 1972): 487–504, and Frank Ervin, "Biological Intervention Technologies and Social Control," *American Behavioral Scientist* 18, no. 5 (May/June 1975): 617–32.

3. Mary Douglas, *How Institutions Think* (Syracuse, NY: Syracuse University Press, 1986), pp. 63, 92.

4. Andrew Abbott has described the role of diagnosis in expanding professional jurisdictions in *The System of Professions* (Chicago: University of Chicago Press, 1988).

5. Diane Chapman Walsh writes about company doctors as double agents in *Corporate Physicians* (New Haven, CT: Yale University Press, 1987).

6. The study was organized by the Genetics Study Group in Cambridge, Mass. Personal communication, 1988–89, Paul Billings, M.D., Ph.D.

7. Robert Blank, "Science and the Human Genome," paper presented at the Conference of the Society for the Social Studies of Science, November 16–19, 1988; and Barbara Katz Rothman, *The Tentative Pregnancy* (New York: Viking, 1986).

8. For a discussion on labeling theory, see Jane Murphy, "Psychiatric Labeling in Cross-cultural Perspective," *Science* 191 (March 1976): 1019–28, and T. Scheff, ed., *Labeling Madness* (Englewood Cliffs, NJ: Prentice-Hall, 1975).

9. Kathleen Nolan and Sara Swenson, "New Tools, New Dilemmas: Genetic Frontiers," *Hastings Center Report* 18 (October/November, 1988): 40–46.

10. *Tarasoff v. Regents of the University of California*, 13 CAL 3d 177, 529 p.2d 553, 118 Cal Reptr 129 (1974).

11. The Social Issues Committee of the American Society of Human Genetics held a workshop on genetic discrimination on November 2, 1986. Reported in Peter Rowley, "Genetic Discrimination," *American Society of Human Genetics* 43 (July 1988): 105–6.

Philip Reilly, "Legal Issues in DNA Banking: Tentative Reflections," draft paper for the American Society of Human Genetics, 1988; see also Neil Holtzman, *Proceed with Caution: The Use of Recombinant DNA Technology for Genetic Testing* (Baltimore: Johns Hopkins University Press, 1989).

12. David Dickson, "Genome Project Gets Rough Ride in Europe," *Science* 241 (February 2, 1989): 599.

INDEX

DATE			